MICRO-COMPUTERS AND AVIATION

AND AVIATION

Paul Garrison

Microcomputers and Aviation

Microcomputers and Aviation

Paul Garrison

A Wiley Press Book
John Wiley & Sons, Inc.
New York • Chichester • Brisbane • Toronto • Singapore

Publisher: Stephen Kippur
Editor: Theron Shreve
Managing Editor: Katherine Schowalter
Composition & Make-up: Publisher's Network

TL
553
G37
1985

Library of Congress Cataloging-in-Publication Data

Garrison, Paul.
  Microcomputers and aviation.

  "A Wiley Press book."
  Includes index.
  1. Aeronautics—Data processing.   2. Aeronautics—
Computer programs.   3. Microcomputers.   I. Title.
TL553.G37   1985   629.13′0028′5416   85-12380
ISBN 0-471-83182-4

Printed in the United States of America

85 86 10 9 8 7 6 5 4 3 2 1

Table of Contents

Introduction   **ix**

PART ONE   **FLIGHT SIMULATOR II**   **xi**

Chapter 1   **The Flight Simulator II Software Package 1**

Chapter 2   **Flight Simulator II   3**

Chapter 3   **First Flight—A Demonstration   10**

Chapter 4   **Sightseeing Above Manhattan   16**

Chapter 5   **A Closer Look at the Editing Menu   21**

Chapter 6   **The Radar View   25**

Chapter 7   **Reading Aviation Charts   28**

Chapter 8   **VOR Departure and Return   42**

Chapter 9   **Cross-Country—New York to Boston   47**

Chapter 10   **Cross-Country—Boston to New York IFR   54**

87- 1822

Chapter 11   **A Look at Approach Techniques   58**

Chapter 12   **From Van Nuys to San Diego
              with Wind and ADF   73**

Chapter 13   **San Diego to Van Nuys After Dark   78**

Chapter 14   **The Angle of Attack   81**

Chapter 15   **Chicago to Seattle in a Hurry   85**

Chapter 16   **Minimum En Route Altitudes
              and Other Matters   91**

Chapter 17   **The War Game   95**

PART TWO   FLIGHT SIMULATOR II   99

Chapter 18   **Commercial Aviation Software—An
              Overview   101**

Chapter 19   **Navlog—A Flight-Planning Program   102**

Chapter 20   **Using Jerry Kennedy's Flight Planner   106**

Chapter 21   **A Family of Aviation Software
              from Flight Opps   110**

Chapter 22   **The RNAV3 Navigator From Briley   114**

Chapter 23   **The A-3 Airplane Simulator   116**

Chapter 24   **The Air Nav Workshop   120**

Chapter 25   **The Ranchele Micro Flight Plan   122**

Chapter 26   **The G-WHIZ Computerised
Flight Planning System   129**

Chapter 27   **Software from PHH Aviation Systems   131**

Chapter 28   **The Dow-4 Gazelle   141**

Chapter 29   **Compuflight Operations Service   142**

PART THREE   DO-IT-YOURSELF AVIATION PROGRAMS   145

Chapter 30   **Do-It-Yourself Programs   146**

Chapter 31   **Weight and Balance for Any Aircraft   149**

Chapter 32   **A Program That Computes
Take-Off Distances   155**

Chapter 33   **Determining the Best Altitude for Cruise   160**

Chapter 34   **Twelve Flight Data Programs   169**

Chapter 35   **Cruise Performance Data for
Piston-Engine Aircraft   174**

Chapter 36   **Flight Route File Program   178**

Chapter 37   **Great Circle Navigation   184**

Chapter 38   **Direct Route Time and Fuel Data   189**

Chapter 39   **Density Altitude   193**

Chapter 40   **Converting Local Time to
Greenwich Mean Time   196**

Chapter 41   **Determining the Advantages of
                  Aircraft Ownership   199**

Chapter 42   **Aircraft Expense Record Program   206**

Chapter 43   **Travel Mode Comparison   215**

Chapter 44   **Converting Local Time to
                  Worldwide Time Zones   221**

Chapter 45   **Exchange Rates for Foreign Currencies   226**

Appendix A   **Converting BASIC-80 to
                  Other Dialects of Basic   229**

Appendix B   **Glossary of Aviation Terms
                  and Abbreviations   233**

Appendix C   **Glossary of Computer Terms
                  and Abbreviations   252**

Appendix D   **Size of the Do-It-Yourself Programs   263**

Appendix E   **Software Manufacturers   264**

Index   267

Introduction

There was a time, not too many years ago, when the word "computer," when spoken to pilots referred to something known for decades as the E6b aviation computer. The E6b is a circular slide rule that can be used to determine range, endurance and many other flight-related data. Some time during the '70s a number of companies, including Jeppesen, Navtronic, Texas Instruments and Hewlett Packard, introduced a family of so-called aviation computers that were really hand-held calculators; they came equipped with aviation programs that performed all the functions of the venerable E6b faster and with greater precision, but also a variety of other calculations important to pilots.

With the proliferation of personal computers in homes and offices, the word computer has become synonymous even in aviation circles with microcomputers, like the Apple II Plus on which this book was written.

In this book I deal with the many different ways in which microcomputers can be immensely useful to pilots and others involved in different phases of aviation. The book is divided into three major parts:

Part One deals with one of the most popular and successful software aviation packages ever devised, the *Flight Simulator II*™, of which some 250,000 have been sold to date. This program, available for the Apple, the IBM PC, the Commodore and several other computers, turns your machine into a fully-equipped stationary single-engine airplane in which you can take off, fly around and land in much the same manner that you would in a real airplane.

Part II concerns itself with a variety of commercially marketed software packages that can be used by pilots for the purpose of flight planning and, to a lesser degree, instrument training.

Part Three is a library of aviation programs that you can key into your own computer (or purchase on disk from the publisher) to perform all manner of tasks, such as determining weight-and-balance conditions, take-off distances under different wind, weather and runway conditions, the best cruising altitude based on winds aloft, and many others.

As I wrote this book I assumed that you are a pilot or, if you are not, are at least reasonably familiar with aviation and its terminology. To assist those readers who may not be sure about the specific meaning of one phrase or another, I have included a glossary of aviation terms, acronyms and abbreviations in the back of the book.

One personal note. You will notice that I use the word *data* as a plural, because I dislike the manner in which such words as data and media are used as singular nouns by many writers and radio and TV news persons.

Paul Garrison
Santa Fe, NM, April, 1985

FLIGHT
SIMULATOR II

The Flight
Simulator II
Software Package

When you purchase *Flight Simulator II*, you not only receive a disk with instructions telling you how to run it, you also receive two fairly well-organized books, two two-sided charts depicting the airports and nav aids in four different areas of the U.S., plus a quick-reference card. The card can be placed next to the keyboard of your computer and used as a reminder of the location of the keys you will need during simulated flight.

The program comes on a 5 1/4-inch disk that requires no preparation or any operating system other than the one built into your computer. The Apple version of the program comes with directions for the Apple II™ or II Plus™ without language card (48K), and with Language Card (64K). The 48K version eliminates certain features, such as the use of an automatic direction finder (ADF), that are available on the 64K version. The program can be run with either a monochrome or a color monitor (I find the color version somewhat more exciting).

The two books are entitled *Pilots Operating Handbook and Airplane Flight Manual* and *Flight Physics & Aircraft Control*, which has an introduction to aerobatics. Both are sturdy 5 1/4 by 7 1/4-inch, 92-page paperback books. The first describes clearly what the program does and all the maneuvers, weather conditions, etc. that can be simulated with it. Also included is an appendix describing the aircraft specifications for the Piper Cherokee Archer II (PA-28-181), the aircraft that is being simulated by the program. You may not wish to read the entire book before firing up the program, but eventually you'll have to read it through if you intend to make full use of everything the program has to offer.

The second book, *Flight Physics & Aircraft Control*, consists of three parts. The first describes why and how an airplane is able to fly in the first place—the physics and aerodynamics of flight—plus a brief and nicely illustrated section describing basic flight maneuvers. This is essential for non-pilots but can be skipped by pilots. The second part consists of eight flight lessons, starting with some very easy ones and ending with fairly difficult instrument approaches. Most of these lessons are well-thought-out and I urge you to fly each one several times. I will not deal with these lessons in this book. The third part is an introduction to aerobatics. Frankly, I believe that attempting to fly aerobatics on this simulator has little practical value. Most aerobatic maneuvers require pilot response to the "feel" of the control surfaces, something that cannot be simulated on a microcomputer.

The two-sided quick reference card shows all the keystrokes and combinations of keystrokes needed to control the aircraft, tune the radios, present differing views out of the cockpit of the aircraft, and so on. You'll probably use this card regularly during simulated flight.

The charts represent four areas, New York/Boston, Los Angeles/San Diego, Chicago/Champaign and Seattle/Olympia. A total of 80 airports and all their associated nav aids are shown on these charts. Except for bodies of water, the charts show no topographical features. Their primary use is to enable the user to quickly locate his or her aircraft on any of the airports by using sets of two coordinates. They're also useful in navigating from one airport to another, although the use of real aviation charts is preferred for that purpose.

Flight Simulator II

The enormous popularity of *Flight Simulator II* may be traced to the fact that it is both a game and a relatively realistic aircraft simulator. Pilots and non-pilots alike can enjoy unlimited hours of exciting and often exacting entertainment or training. The program is a masterpiece of excellent graphics. The screen is divided into two halves; the upper half depicts the outside world, and the lower half shows an instrument panel. The screen is a reasonably close representation of the average single-engine light aircraft instrument layout (Figure 2.1).

Since a clear understanding of the instruments and their functions is a prerequisite for making any meaningful use of the program, let's talk about them first. In the upper left-hand corner of the instrument panel is the airspeed indicator, calibrated for speeds from 40 to 160 knots. Below it is the turn coordinator, representing the bank angle of the aircraft in the form of a rotating line, and yaw, if any, in the form of a slip indicator or ball.

To the right of the airspeed indicator is the artificial horizon (also referred to as attitude indicator), which displays the attitude of the aircraft with reference to the horizon. To the right of this instrument in the top row is the altimeter, equipped with a setting knob that is used to adjust the displayed altitude to the current barometric pressure. Between the two, at the very top of the instrument panel, is the stall warning light, which lights up when the aircraft is about to lose flying speed.

Below the artificial horizon is the directional gyro (DG or heading indicator), equipped with a knob used to adjust its indication to that of the

Figure 2.1 The upper half of the display shows the view seen through the windshield; the lower half is a realistic representation of the instrument panel.

magnetic compass. To its right is the vertical speed indicator, calibrated from zero (level flight) to 1,500 feet per minute (fpm) climb or descent. Between the two is a three-part instrument not normally part of the average aircraft instrument panel. Its top line indicates the position of the ailerons, the vertical line in the center indicates the position of the elevator, and the horizontal line at the bottom indicates the position of the rudder. These unconventional indicators are included on the simulator panel because the *keyboard* must be used to adjust these control surface positions, thereby, eliminating the "feel" the pilot normally depends on when flying an aircraft.

That completes the basic instrument cluster. To the right of it are the navigation avionics and the engine instruments, fuel gauges, etc.

Directly to the right of the basic instrument cluster are two omnibearing indicators (OBI). The upper one includes a glide-slope indicator. Both are equipped with TO/FR indicator and a knob to adjust the course indicator. Directly to the right, just under the top edge of the instrument panel, is a vertical rectangle that represents the flap position indicator.

Figure 2.2 Here is another forward view, while the RPM shows the engine at full throttle, the airspeed indicator shows 78 knots, and the vertical speed indicator shows the beginning of lift-off.

Below it, in the center, is a similar-appearing instrument representing the trim position indicator. And below that, at the very bottom, is a trapezoidal shape containing the throttle-setting indication, marked T, and the mixture indicator, marked M. In this row atop the edge of the instrument panel is the magnetic compass.

To the far right, the top half of the instrument panel is divided into six rectangles representing the radios. At the top are the communication radio (COM 1) and the outer, middle and inner marker lights (O M I). Below them is navigation radio number one (NAV 1) and the distance measuring equipment (DME). In the next row are navigation radio number two (NAV 2) and the transponder (XPNDR).

The bottom half of that section contains a digital 24-hour clock, magneto selector/indicator, lights on/off indicator, fuel gauges for the right and left tank, oil temperature and pressure indicators, the rpm indicator, carburetor heat indicator and suction gauge.

If you've never actually watched this program in operation, you are probably skeptical about the possible realism of the instrument indications

Figure 2.3 Here the view is a 45-degree angle to the left, with the left wing showing in the left bottom corner of the scene.

during simulated flight. In fact, these instruments function properly with an absolute minimum of compromise. For instance, if you run one of your fuel tanks dry the engine quits.

Whenever a pilot gets into an unfamiliar airplane, he will want to sit in the pilot's seat and study the instrument panel for a while, to become familiar with the placement of the instruments. If you're unfamiliar with *Flight Simulator II*, boot the disk and study the representation of the instrument panel for a while. While doing so you might want to reread the descriptions above, or open the *Pilot's Operating Handbook* and *Airplane Flight Manual* to pages 14 and 15, where you'll find a schematic representation of the instrument panel along with a brief description of each. You will find this time well spent, since it may help you avoid all sorts of annoying errors later on.

So far we've talked only about the bottom half of the display on the screen, the instrument panel. Let's take a closer look at the top half. In its default condition this half represents a straight-ahead view of the wind-

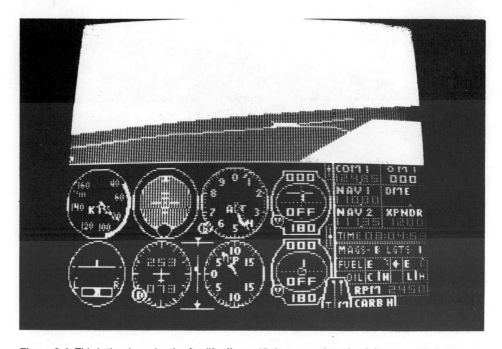

Figure 2.4 This is the view, shortly after lift-off, at a 45-degree angle to the right.

shield. The program has the capability to look in other directions—actually more than are available in a real airplane. The available views are:

1.   straight ahead
2.   45 degrees to either left or right
3.   90 degrees to left and right
4.   135 degrees to left and right
5.   straight back (180 degrees)
6.   straight down, as if through the floor of the aircraft.

The scenery changes realistically as we look in different directions, and if we happen to be directly over an airport, we can look down and see the runway design (see Figures 2.2 through 2.6).

Since the program uses a great many keystrokes and combinations of

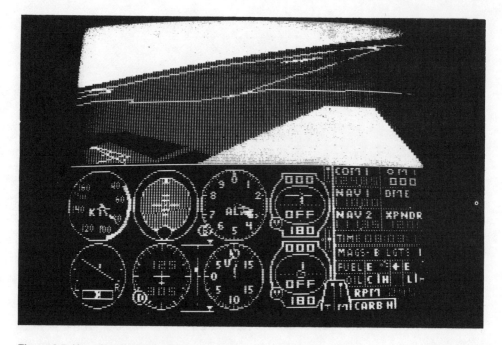

Figure 2.5 Here the view is at a 90-degree angle to the left with the aircraft in a bank and the airport runway pattern below.

keystrokes to control the flight, adjust the instruments, and so on, you'll want to keep the quick reference card handy until you become thoroughly familiar with which keystroke produces what result.

Having gotten this far, let's take our first flight.

Figure 2.6 This view is at a 135-degree angle looking back to the left, with the vertical stabilizer seen to the left.

First Flight
A Demonstration

To discover what *Flight Simulator II* is all about, we'll start by taking a demonstration ride. Boot the disk in the normal manner and wait ten seconds or so while the disk drive clacks and whirs. Eventually an incomplete representation of the instrument panel will appear, and above it a legend will announce that the computer is loading the program into RAM. Shortly thereafter that legend changes to another, asking whether you're using a color or black and white monitor (Figure 3.1). The program functions equally well with either, though the full beauty of the graphics can only be appreciated in color.

Once you have typed A or B to reply to this question, the legend changes again, now asking whether you want the demo flight or a regular flight (Figure 3.2). Since this is your first exposure to the program, select the demo flight. This produces the information that the demo flight runs continuously, and that you can exit it to take over the controls by typing K (Figure 3.3). For the time being press any key except K to start the flight.

Now sit back and relax. Pretend you're sitting in the right seat of the aircraft with an experienced pilot in the left seat flying the airplane. It takes a few seconds while the display is organized before you find yourself sitting on the ground at Meigs Field in Chicago on runway 36, facing north. As power is applied (watch the throttle indicator and the rpm numbers), the instruments start to display the appropriate data—such as field elevation (592 feet) on the altimeter, the direction in which the aircraft is facing on the magnetic compass and the directional gyro, and so on. The airplane taxis into position on the center line of the runway. In the distance to the left you see downtown Chicago; the John Hancock Building is clearly visible.

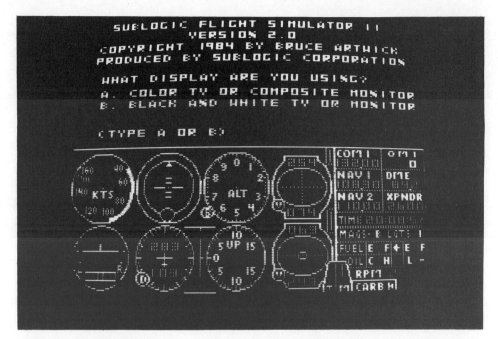

Figure 3.1 After the program has been loaded, you're asked whether you're using a color or black and white monitor.

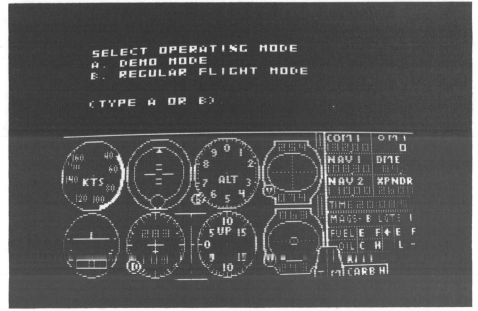

Figure 3.2 Next you're asked to decide on a flight mode.

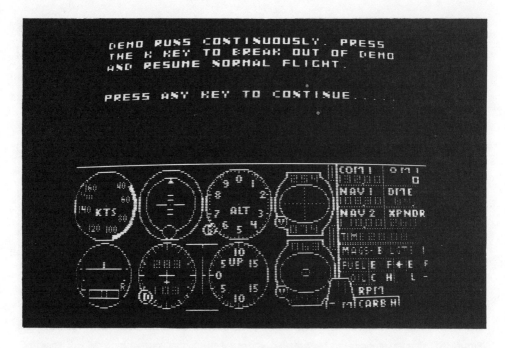

Figure 3.3 If you select the demo flight, you're informed that in this mode the flight continues indefinitely and you must use the K key to switch to normal flight.

The airplane starts its take-off run. As the needle on the airspeed indicator approaches 60 knots, the nose of the aircraft rises and it begins to lift and climb. The vertical speed indicator shows the rate of climb, and the altimeter needle indicates the increase in altitude, while the artificial horizon and the turn-and-bank indicator show that you're continuing the climb in a wings-level attitude (Figure 3.4).

As the John Hancock Building is about to disappear from view, press the 5 key followed by the R key, and you'll be looking out to the left at a 45-degree angle, seeing part of the left wing and the building slowly disappearing under it. If you press 5 followed by V, you'll be looking back at a 135-degree angle, watching the building reappearing under the aft edge of the wing and disappearing in the distance. Since someone else is flying the airplane, you might want to experiment with the various views available by pressing 5 followed by the different keys that control the view (see Figure 3.5).

By now the airplane has begun to bank. If you're back to the view facing front (5 T), you'll be able to compare the angle of the horizon with

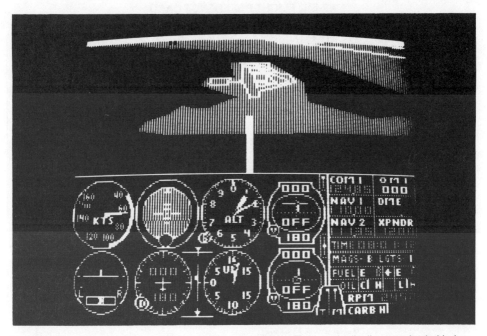

Figure 3.4 Looking out the rear of the airplane, we see Meigs Field gradually disappearing behind us.

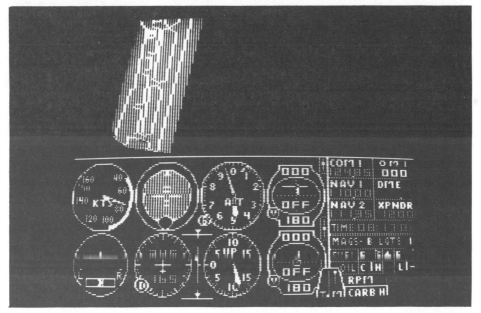

Figure 3.5 When overflying the field, we can look right through the floor of the airplane at the airport below us.

A2-FS2 FLIGHT REFERENCE CARD

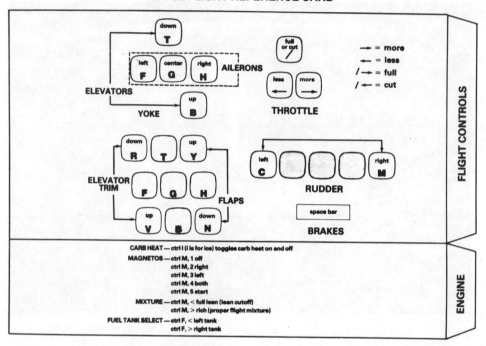

Figure 3.6 The reference card showing the keys that control throttle, ailerons, elevator, rudder and brakes. (Courtesty SubLogic).

the indication of the artificial horizon and that of the turn-and-bank indicator. The airplane will perform a variety of maneuvers, some of which seem a bit extreme as it climbs sharply while dissipating airspeed and then descends to regain speed. As mentioned before, this flight is a loop that runs continuously, so use this opportunity to learn to scan the instrument cluster and to become accustomed to recognizing what the various indicators are telling you. For the time being, ignore the radio stack and the other instruments on the right third of the panel and concentrate on the six primary flight instruments. I would suggest that you keep this up for at least 15 or 20 minutes, even if you're a rated pilot, because it takes at least that long to become comfortable with the instrument panel.

If at any time you want to look at the current attitude of the aircraft and the resulting instrument indications at leisure, press the P key (P for Pause) and everything will stop. The P key is a *toggle key*, meaning that pressing it a second time will cause the flight to continue.

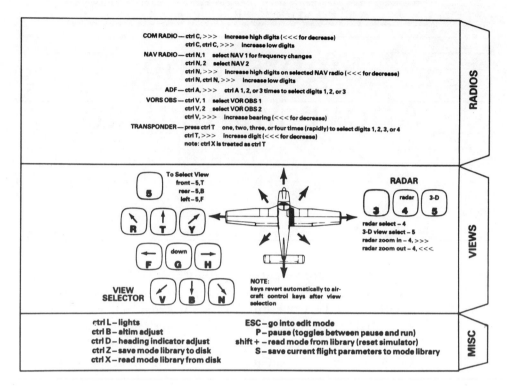

Figure 3.7 The reverse side of the reference card, showing the views and the keys used to call them from the keyboard. (Courtesy SubLogic).

When you've had enough, you'll be faced with one of the peculiarities of the program that I consider a drawback: there's no way to quit. If you want to quit now, make sure the disk drive light is not lit, indicating that the drive is not doing anything, then remove the disk and turn the computer off. If you'd rather try your hand at flying the airplane, press K and you'll be in control. With the flight reference card (see Figures 3.6 and 3.7) in front of you, operate the ailerons to control the lateral attitude of the aircraft (keys F G and H), and the elevators to control the pitch attitude (keys T and B), and observe what happens. If you crash (producing the word CRASH on the screen), don't worry. The display will simply return to the beginning of the flight.

I suggest you practice for a half an hour or so. By that time you'll find you're beginning to get tired and that your attention span is poor. Quit and do something else, taking your next flight when you're rested and ready to go.

Sightseeing
Above Manhattan

Let's take some time for a sightseeing flight around Manhattan. This is something you can only do in the simulator, as the airspace over and around Manhattan is a TCA (Terminal Control Area), where all aircraft must comply with directions issued by the ATC (Air Traffic Control).

Before starting we must position our airplane in one of the airports in the New York area. Let's take LaGuardia Airport. Look at the New York and Boston Area Chart that is part of your *Flight Simulator II* package; you'll find LaGuardia in the left bottom corner. Now look at the Airport Directory at the bottom of that chart, and find the coordinates for LaGuardia (17091 north and 21026 east). The airport is at an elevation of 22 feet. You'll use these data and the fact that you want to use Runway 31 for take off to position your aircraft at (or near) LaGuardia Airport. (The north and east coordinates have no relationship to actual latitude and longitude figures; LaGuardia is actually located at 40-46 lat. and 73-52 long.)

Boot the disk and select the regular flight from the menu. When the display shows you on the ground at Meigs Field in Chicago (as during the demonstration flight), press the Escape key and the display changes to a large menu used to change any of the default values (Figure 4.1). Press the Return key repeatedly until the pointer is opposite the north position category. Type 17091 and press the Return key to place the pointer opposite the east position; type 21026. Next place the pointer opposite the altitude category and type 22. Finally, place the pointer opposite the heading and type 310, the direction in degrees of runway 31.

Once these data have been entered, press the Escape key. After a few moments (during which the program gets its act together), you'll find

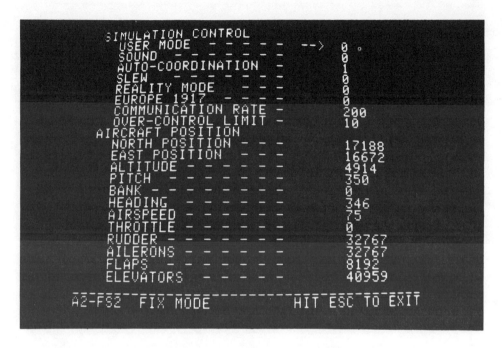

```
SIMULATION CONTROL
  USER MODE   - - - - -    -->    0 °
  SOUND   - - - - - - -            0
  AUTO-COORDINATION   -            1
  SLEW   - - - - - - -             0
  REALITY MODE   - - - -           0
  EUROPE 1917   - - - -            0
  COMMUNICATION RATE  -          200
  OVER-CONTROL LIMIT  -           10
AIRCRAFT POSITION
  NORTH POSITION  - - -        17188
  EAST POSITION   - - -        16672
  ALTITUDE - - - - - - -        4914
  PITCH   - - - - - - -          350
  BANK  - - - - - - - -            0
  HEADING   - - - - - -          346
  AIRSPEED  - - - - - -           75
  THROTTLE  - - - - - -            0
  RUDDER  - - - - - - -        32767
  AILERONS  - - - - - -        32767
  FLAPS   - - - - - - -         8192
  ELEVATORS   - - - - -        40959

A2-FS2   FIX MODE            HIT ESC TO EXIT
```

Figure 4.1 The menu that permits you to select a variety of simulation modes.

yourself on the ground, surrounded by mostly dark areas which you can assume represent the runway at LaGuardia. No action will be taking place except for the digital clock which will be counting seconds and minutes; your engine has not been started. Since we have not yet advanced to the more sophisticated flight parameters, we can skip magneto checks and other preflight adjustments. Simply press the Right Arrow key repeatedly (or press / followed by the Right Arrow key) and watch the throttle indicator move to the top of the vertical line. The rpm will go up to the full-throttle setting, and the aircraft will start to move. As the speed shown on the airspeed indicator approaches lift off speed, somewhere around 50 knots, start to pull back on the yoke slowly (press the B key two or three times), and after a moment or so you'll see the ground fall away and the vertical speed indicator showing the rate of climb.

Be sure to adjust your rate of climb to an acceptable rate of 500 feet per minute (fpm) by adjusting the elevator position. Keep climbing straight ahead until you've reached a safe altitutde, say, 1,000 feet (check the altimeter) before initiating any turns. Once at or above 1,000 feet, you might enter a shallow bank to the left by pressing F once. When the

Figure 4.2 Approaching Manhattan, we can distinguish the Empire State Building to the left and the twin-tower World Trade Center in the distance.

appropriate bank angle has been reached, press G to neutralize the ailerons again. You'll still be climbing and after a moment you'll see the shape of Manhattan Island along with streets and two tall buildings representing the Empire State Building and the twin towers of the World Trade Center. Stop the turn by typing H followed by G once the turn-and-bank indicator shows a wings-level attitude (Figure 4.2).

By now you should have reached about 2,000 altitude, it's time to level off. Reduce power by pressing the left arrow key several times until the rpm indicator is around 1950, which should result in level flight with no need for any other adjustments (check your vertical speed indicator). If you've turned just the right amount, you should be flying straight toward the center of Manhattan and those buildings. If not, make whatever heading adjustments are necessary to stay on course toward those buildings. Once you pass the Empire State Building (if you're too low you'll crash into it), you might want to look out the side or back of the airplane to see it moving away from you.

When you again look out the front, the World Trade Center will be

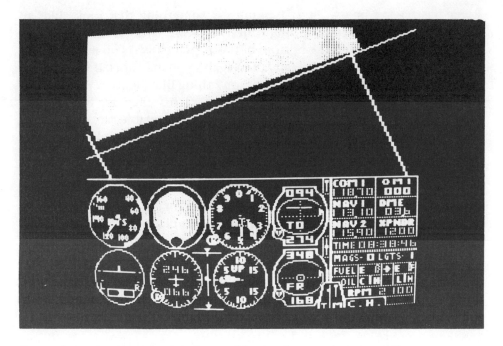

Figure 4.3 Looking down through the bottom of the aircraft, we see the runway pattern.

visible and in the distance is a tiny spot in the Hudson River that turns into the Statue of Liberty as you approach. Continue to fly in that direction until you reach it. You might then want to turn 180 degrees and retrace your steps. I suggest flying around Manhattan for a while in this manner to get the feel of banking and turning. Learn not to overcontrol, something we're tempted to do especially when we've inadvertantly gotten ourselves into an extremely steep bank.

Remember, when banking, the aircraft develops less lift than in level flight and you may have to apply back pressure to the elevators (B key) or increase the throttle setting by a few notches (Right Arrow key).

Eventually you'll want to get back to LaGuardia Airport. Since on this flight you're not using navigation radios, you'll have to do it by pilotage (dead reckoning). The approximate heading from the Empire State Building to LaGuardia is in the neighborhood of 95 or 100 degrees. Once you're near that building, take up such a heading. Be sure to check your heading by the directional gyro (DG) rather than the magnetic compass, because magnetic compass readings are unreliable during turns. When the East

River has passed below, start looking for something that looks like an airport. After what seems a long time, you'll be flying over some dark patches. If you look out the bottom of the airplane (5 G) you may actually see the runways at the airport (and then again, you may not). If you can see the runways your screen should look something like Figure 4.3.

Don't worry about landing on a runway at this stage; we'll get to that later. For the time being, simply try to set up a reasonable rate of descent, something like 500 fpm, by reducing the throttle setting and slowly raising the elevator. This takes practice! Watch the vertical speed indicator, the airspeed indicator and the altimeter. You should be gliding down at about 80 knots with the throttle all the way or nearly all the way back. You'll probably crash the first dozen or so times that you try to land (I did). This phase of flight is particularly difficult on this simulator, because, unlike in a real airplane (or a real simulator with visuals), it is virtually impossible to gauge the distance between the airplane and the ground.

So far we've flown without radios and without using any of the available navigation aids. We'll take a look at those in some of the next chapters.

A Closer Look at the Editing Menu

The two-page editing menu can be difficult. It is not easy to remember what to type in reply to the different subjects to achieve a given result. In this chapter I'll try to provide a quick reference guide that should solve the problem. You might want to photocopy these pages to have them handy whenever you start on a new flight or maneuver.

Simulation Control

USER MODE—the default setting is 0. The available options are:

0   Easy flight

1   Realistic fair-weather flight

2   Self-flight demonstration

3   Dusk flight

4   Night flight

5   Moderate weather flight

6   Bad weather flight

7   WW I Ace aerial battle game

8   Preset flight mode

9   Preset flight mode

10 – 24 Available for user-designed modes.

To find out what takes place when these modes are selected, you might want to experiment by making short flights using each in turn.

SOUND—the default is 0, meaning no sound. Typing 1 produces some minor sound effects.

AUTO COORDINATION—the default is 1, producing the effect of aileron/rudder linking. Type 0 and rudder and ailerons must be individually controlled.

SLEW—the default is 0, or off. Typing 1 turns it on, permitting extremely fast airspeeds in order to quickly move to some distance destination.

REALITY MODE—the default is 0 or off. Typing 1 turns this mode on, requiring greater precision in operating the aircraft.

EUROPE 1917—the default is 0 for normal flight simulation. Typing 1 produces the WW I aerial battle game.

COMMUNICATION RATE—the default is 200. This parameter controls the speed with which communication messages appear on the screen. 200 is much too fast; try 20. You have a choice from 1 (slow) to 255 (fast).

OVER CONTROL LIMIT—the default is 10. This is not explained anywhere in the handbook; you might try to experiment with other values.

Aircraft Position

NORTH POSITION—the default is 17189 for Meigs Field

EAST POSITION—the default is 16671 for Meigs Field

ALTITUDE—the default is 592 for Meigs Field

PITCH—the default is 0. Best leave it alone.

BANK—the default is 0. Leave it alone too.

HEADING—the default is 0, or due north. Use any number from 0 – 359 to control the heading of the aircraft when the edit mode is exited.

AIRSPEED—the default is 0. You may enter any airspeed in knots from 0 to 154, the never-exceed speed of the simulated aircraft.

THROTTLE—the default is 0. The available range is from 0 (idle cut off) to 32767 (full throttle). These numbers have no direct relationship to the rpm.

RUDDER—the default is 32767 for centered. The available range is from 1024 – 64512.

AILERONS—the default is 32767 for neutral. The available range is from 1024 – 64512.

FLAPS—the default is 0, no flaps. The available range is from 0 –24576.

ELEVATOR—the default is 32767, for neutral. The available range is 12288 – 53248.

Note: Unless you want to find yourself in some unusual attitude when you're returning to the program, I suggest leaving rudder, ailerons, flaps and elevator in the default settings.

Environmental Control

TIME: HOURS—the default is 8. Enter any hour, using the 24-hour clock.
       MINUTES—the default is 0. Enter any value from 0 – 59.

SEASON—the default is 4 for fall. 1 = winter, 2 = spring and 3 = summer.

CLOUD LAYER 2: TOPS—the default is 0. Enter any altitude.
               BOTTOMS—the default is 0. Enter any altitude.

CLOUD LAYER 1: TOPS—the default is 0. Enter any altitude.
               BOTTOMS—the default is 0. Enter any altitude

WIND LEVEL 3: KNOTS—the default is 0.
               DEGREES—the default is zero. Available: 0 – 359

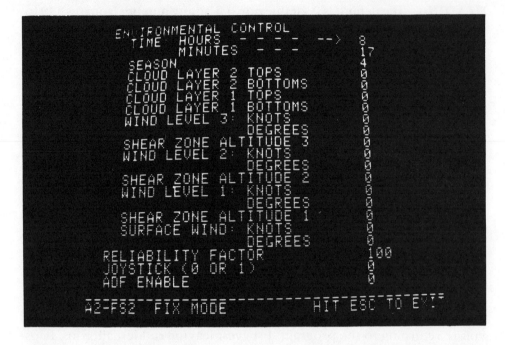

Figure 5.1 The second page of the edit menu is used to select wind and weather conditions and to activate the ADF if required.

SHEAR ZONE ALTITUDE 3—the default is 0. Enter any altitude. Ditto for **WIND LEVELS** and **SHEAR ZONE ALTITUDES** 2 and 1 and for **SURFACE WIND**.

RELIABILITY FACTOR—the default is 100 for no problems, or 100 percent liability. Lower numbers, from 1 up, result in problems or malfunctions with increasing frequency, as the number decreases. Available only in the REALITY mode.

JOYSTICK (0 or 1)—the default is 0, for no joystick.

ADF ENABLE—the default is 0. Typing 1 replaces NAV 2 and OBI 2 with the ADF and its display.

The Radar View

In addition to all those angles at which you can look out the simulated airplane, one more comes in extremely handy when you need to orient yourself and you're not sure exactly where you are in relation to certain ground features. This view is referred to as the "radar view" in the documentation literature that accompanies *Flight Simulator II*. It shows the airplane in the center of the upper portion of the display, along with the available ground features, if any. The view is called up by typing 4. You then use the < and > keys to increase or decrease the viewing area.

To demonstrate, let's once more take the demonstration ride so you can look at what happens without having to worry about controlling the airplane. As soon as the aircraft starts to lift off, type 4 and follow that by two or three strokes of the < key. You'll see a sort of bird's-eye view of your airplane with the airport and surrounding area in the background (Figure 6.1). Be patient and watch what happens. After climbing straight north, the airplane makes a number of rather erratic turns before eventually flying away from the lake and over the city. You'll notice that it is the *background* that turns, not the airplane symbol itself. This is realistic; when you're sitting in a real airplane and fly turns, it is the ground below that appears to be turning in relation to your position in the aircraft.

Once the airplane is over downtown Chicago (represented by little black dots), press the < key two or three times and you'll see O'Hare Airport to the right, and Midway to the left of the nose of the airplane. The airplane turns again and flies toward the departure airport, Meigs Field, but you may have to press the > key a number of times to get a clear view of that airport and its runway.

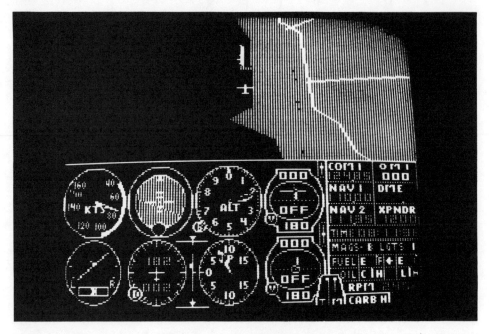

Figure 6.1 The radar view shows your airplane in relation to Meigs Field with downtown Chicago to the right.

I'm sure you have already realized that had we used this radar view while trying to find our way back from Manhattan to LaGuardia Airport in Chapter 4, it would have simplified the task. As a matter of fact, the radar view comes in especially handy during two important phases of flight. One is a cross-country trip, where we want to be able to orient ourselves with reference to the towns, roads, airports and other landmarks shown on the charts. The other is prior to making an approach to a landing when we want to fly a reasonably descent pattern, consisting of the downwind and base legs and the final approach.

Before going much further, it would be a good idea if you made a number of practice flights, possibly repeating the sightseeing flight over Manhattan. This time make good use of the radar view, just as you would in a real airplane. Look at ground-based features and locate your aircraft's position relative to them. This kind of practice will teach you when to roll out of a turn in order not to overshoot the desired new heading. Be sure to get your airplane leveled off at some safe altitude, high enough to see a lot of the ground below (say, 4,500 feet), and have set up level flight that doesn't require a great deal of your attention on the throttle or elevators to maintain the established pitch attitude.

We'll be using the radar view capability liberally in subsequent training flights, under VFR conditions.

Reading
Aviation Charts

Once we start to make some simulated cross-country flights you'll discover it is definitely desirable to use some actual aviation charts rather than those supplied with the program. If you're a pilot you probably have charts for at least some of the four areas for which airport locations and nav aids are stored in the program. If you're a non-pilot, you might be able to get some outdated charts from a pilot friend, or you can obtain them from the FBO at your local airport.

To effectively fly some of the flights described in this book, you should use two types of charts. For flights under VFR conditions use sectionals that show all meaningful topographical features, giving you an opportunity to compare them with those shown by the program. Flying long distances in this manner makes the flight more interesting.

For flights under IFR conditions you'll need *low altitude en route charts*, either those issued by the government and available from FBOs at airports, or those issued by Jeppesen, available only on a subscription basis. Any instrument-rated pilot is likely to have a bunch of old ones kicking around. These charts show no topographical features, but display the nav aids, the airways, distances between fixes, minimum safe altitudes and just about everything else that is needed in order to first plan an instrument flight and then to actually execute it. (Figures 7.1 through 7.12).

Full color sectionals, used with some of the flights described in this book have been reproduced in black and white. The instrument charts are printed in blue and green, with airports and important topographical features shown in green, and the nav aids, airways, and so on shown in blue. Those shown alongside several IFR flights are produced by Jeppesen, and reproduced with permission of that company.

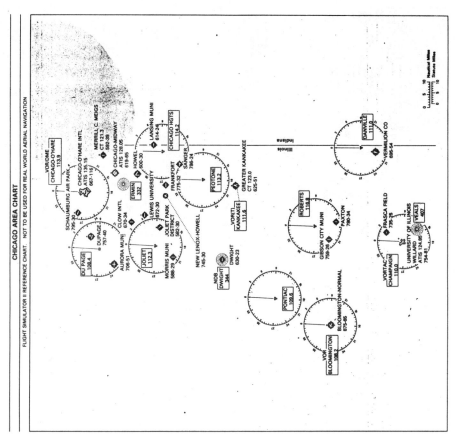

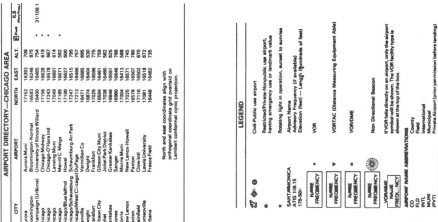

Figure 7.1 The Chicago area chart depicts the available airports and navigation aids. (Courtesy SubLogic).

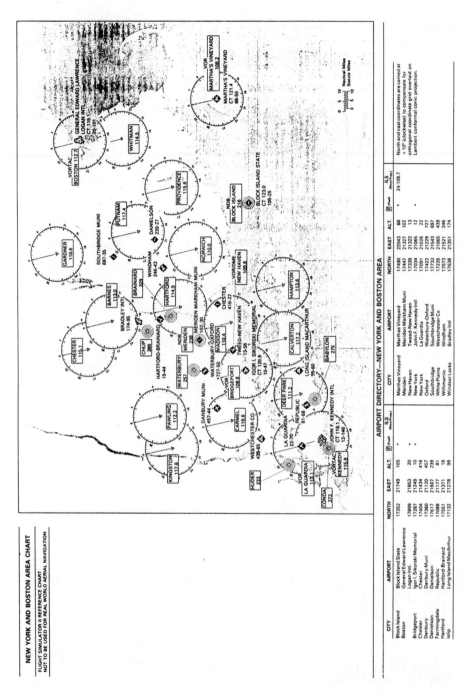

Figure 7.2 The New York and Boston area chart depicts the available airports and navigation aids. (Courtesy SubLogic).

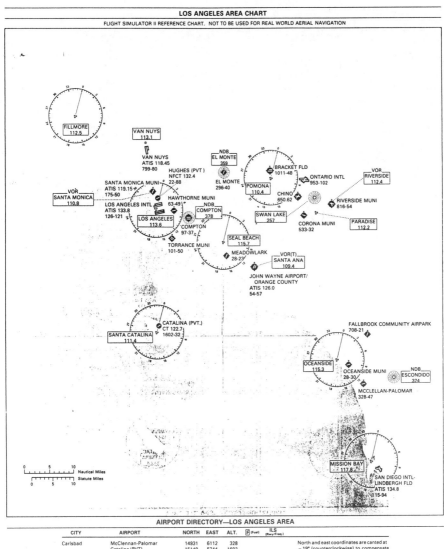

LOS ANGELES AREA CHART

FLIGHT SIMULATOR II REFERENCE CHART. NOT TO BE USED FOR REAL WORLD AERIAL NAVIGATION

AIRPORT DIRECTORY—LOS ANGELES AREA

CITY	AIRPORT	NORTH	EAST	ALT.	F (Fuel)	ILS (Rwy/Freq.)
Carlsbad	McClennan-Palomar	14931	6112	328		North and east coordinates are canted at
	Catalina (PVT)	15149	5744	1602		−19° (counterclockwise) to compensate
Chino	Chino	15319	6079	650		for orthogonal coordinate grid overlaid
Compton	Compton	15,334	5859	97		on Lambert conformal conic projection.
Corona	Corona Muni	15280	6083	533		
El Monte	El Monte	15397	5952	296		
Fallbrook	Fallbrook Community Airpark	15023	6144	708		
Hawthorne	Hawthorne Muni	15358	5831	63		
Huntington Beach	Meadowlark	15244	5911	28		
LaVerne	Brackett Fld	15378	6038	1011		
Los Angeles	Hughes (PVT)	15386	5808	22	*	
Los Angeles	Los Angeles Intl	15374	5805	126	*	
Oceanside	Oceanside Muni	14974	6095	28		
Ontario	Ontario Intl	15347	6099	952		
Riverside	Riverside Muni	15288	6141	816		
San Diego	San Diego Intl–Lingbergh Fld	14761	6102	15	*	
Santa Ana	John Wayne Airport/	15211	5981	54		
Santa Monica	Santa Monica Muni	15402	5799	175	*	
Torrance	Torrance Muni	15308	5815	101		
Van Nuys	Van Nuys	15498	5811	799	*	16R/111.3

Figure 7.3 The Los Angeles area chart depicts the available airports and navigation aids. (Courtesy SubLogic).

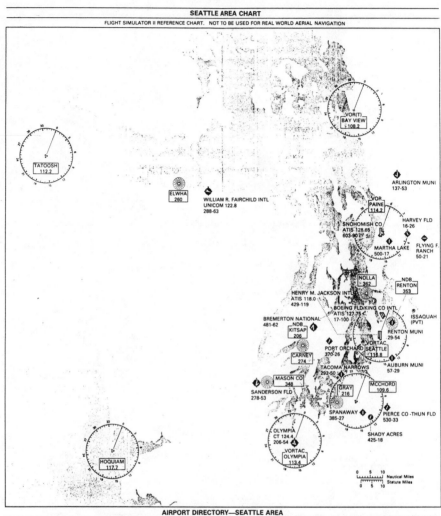

SEATTLE AREA CHART

FLIGHT SIMULATOR II REFERENCE CHART. NOT TO BE USED FOR REAL WORLD AERIAL NAVIGATION

AIRPORT DIRECTORY—SEATTLE AREA

CITY	AIRPORT	NORTH	EAST	ALT.	F (Fuel)	ILS (Rwy/Freq.)
Alderwood Manor	Martha Lake	21502	6670	500		
Arlington	Arlington Muni	21616	6737	137		
Auburn	Auburn Muni	21290	6586	57		
Bremerton	Bremerton National	21407	6470	481		
Everett	Snohomish Co	21525	6665	603	•	16/109.3
Issaquah	Issaquah	21362	6668	500	•	
Monroe	Flying F. Ranch	21481	6738	50		
Olympia	Olympia	21218	6343	206	•	
Puyallup	Pierce Co.-Thun Fld	21206	6534	530		
Port Angeles	William R. Fairchild Intl.	21740	6375	288	•	
Port Orchard	Port Orchard	21373	6483	370		
Renton	Renton Muni	21351	6612	29		
Seattle	Boeing Fld/King Co Intl	21376	6596	17	•	
Seattle	Henry M. Jackson Intl. (Seattle-Tacoma Intl)	21343	6584	429		
Shelton	Sanderson Fld.	21353	6316	278		
Snohomish Co. (Paine Field) see Everett						
Snohomish	Harvey Fld	21505	6711	16		
Spanaway	Shady Acres	21201	6501	425		
Spanaway	Spanaway	21215	6491	385		
Tacoma	Tacoma Narrows	21300	6480	292		

North and east coordinates are canted at −21° (counterclockwise) to compensate for orthogonal coordinate grid overlaid on Lambert conformal conic projection.

Figure 7.4 The Seattle area chart depicts the available airports and navigation aids. (Courtesy SubLogic).

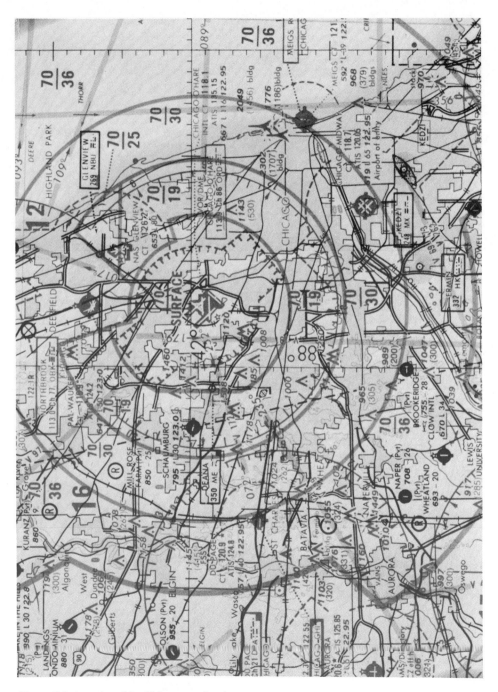

Figure 7.5 A portion of the Chicago sectional.

Figure 7.6 A portion of the New York sectional.

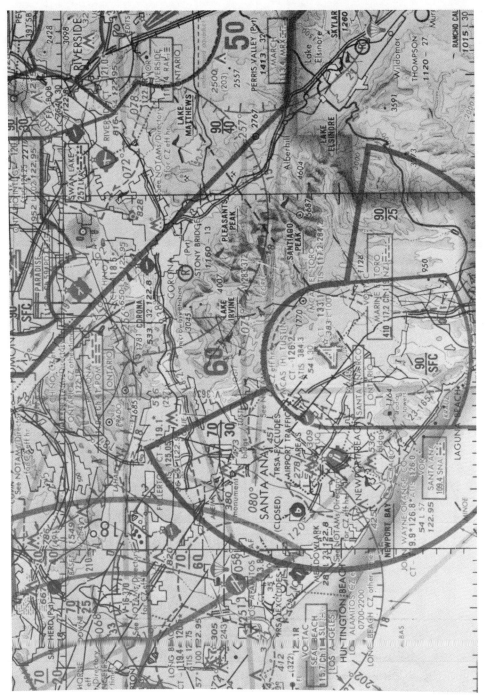

Figure 7.7 A portion of the Los Angeles sectional.

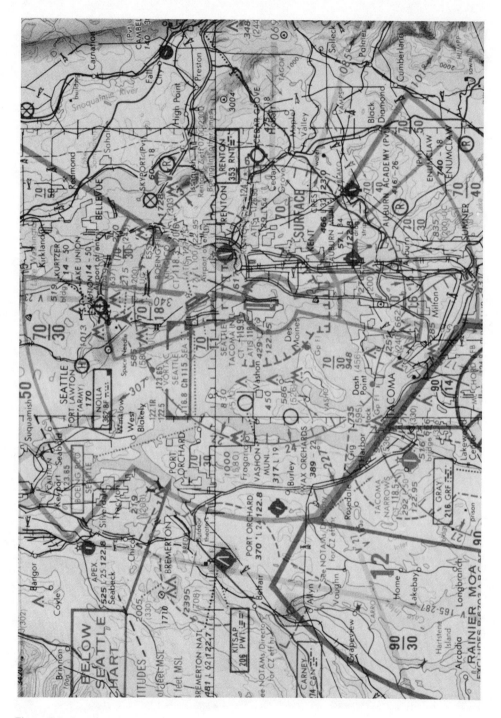

Figure 7.8 A portion of the Seattle sectional.

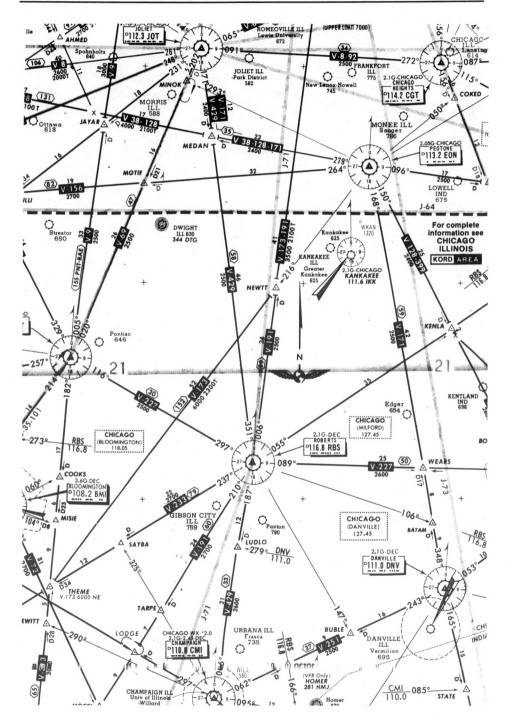

Figure 7.9 A portion of the Chicago area low altitude chart. (Copyright©Jeppesen & Co.).

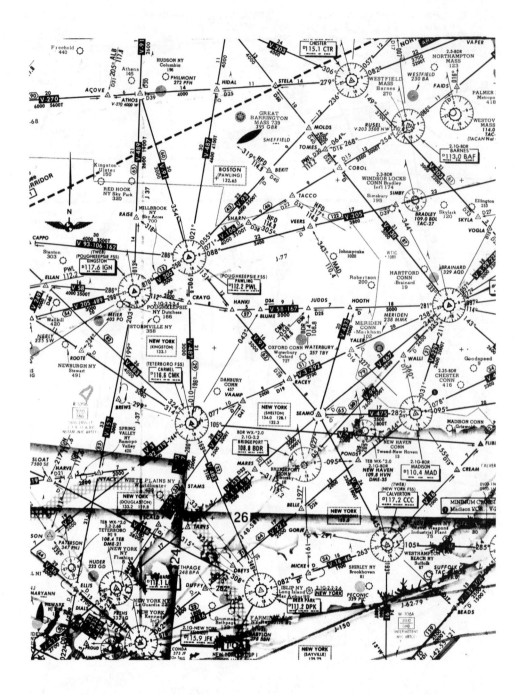

Figure 7.10 A portion of the New York and Boston area low altitude chart. (Copyright© 1985 Jeppeson & Co.).

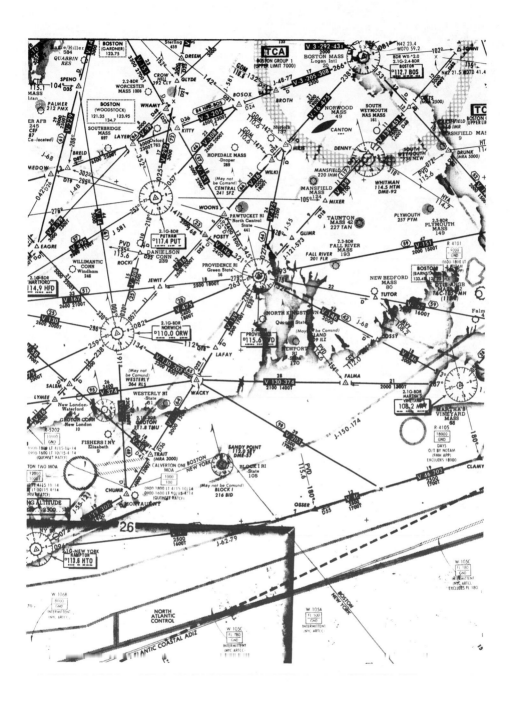

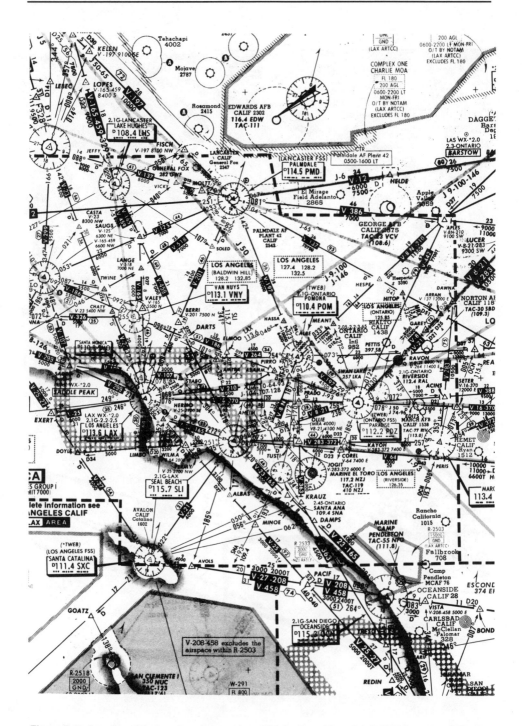

Figure 7.11 A portion of the Los Angeles area low altitude chart. (Copyright© 1985 Jeppeson & Co.).

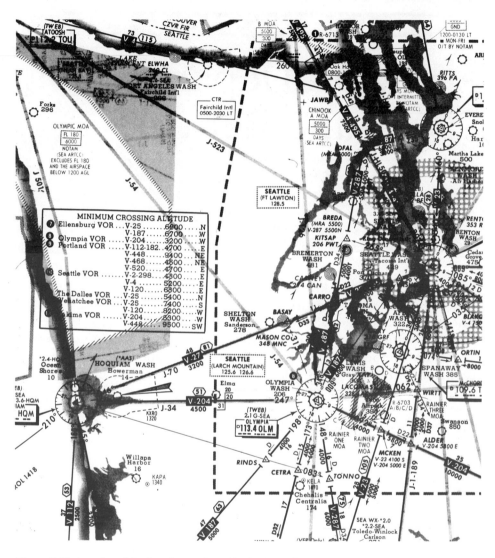

Figure 7.12 A portion of the Seattle area low altitude chart. (Copyright© 1985 Jeppeson & Co.)

Reading aviation charts takes practice. If you're a non-pilot, I'd suggest that you spend a few hours studying them and looking at the descriptions that explain what all those symbols and hieroglyphics mean. A course in reading aviation charts is beyond the scope of this book. If you have trouble, talk to a pilot or flight instructor and ask him or her to help you. Most will be glad to do so.

VOR Departure and Return

Let's learn to taxi and intercept a VOR radial, establish ourselves on the outbound course, then fly a 180-degree turn and return to the airport. The entire maneuver should resemble the pattern shown in Figure 8.1. After activating the program and selecting regular flight, press the ESC key, and with the menu in display and the arrow opposite the User Mode category (top line) type 8 followed by the ESC key. When the display settles down you'll be sitting in your airplane somewhere on Willard Airport in Champaign, Illinois (see the illustration on page 51 of the Flight Physics and Aircraft Control booklet that came with your simulator).

To get a clearer picture of where you're located in the airport, type 4 to call up the radar view, and then press the < key a number of times until the entire airport is in view (Figure 8.2). Check the DG and magnetic compass indicators to determine the direction in which you're facing. They will have settled down on 90 degrees, meaning that you're facing east, east being at the top of your screen, north to the left, west to the bottom and south to the right.

Now check your instrument indicators to make sure the elevators, ailerons etc. are in neutral position. Look at your nav radio. For this exercise we're only concerned with the NAV 1, at the top, tuned to 110.0, the frequency of the Champaign VOR (CMI). This is located roughly half-way between the take-off points of runways 31 and 36, directly south of your current position in the airport. As an exercise, press CTRL N followed by several < and > key presses to see that the frequency displayed by the NAV 1 is selected in this manner. You might also try pressing CTRL

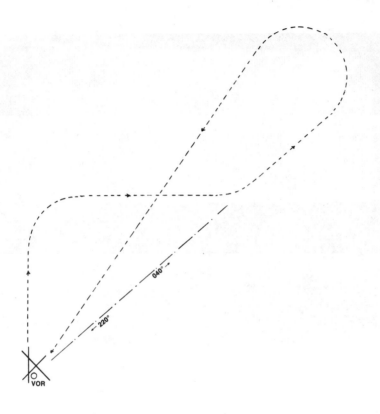

Figure 8.1 The approximate flight path (not to scale) after departing Willard's Runway 36 and then returning to the VOR.

N twice followed by < < < and > > > to observe how to tune the decimal portion of the frequency display.

Next press CTRL V followed by < < < or > > > to change the heading indicator which, at the moment, is displayed as 000. As you slew it through 180 and 360 degrees, you'll see that the needle briefly centers, because you're currently located close to the 360-degree radial from the VOR. Now select the radial you'll want to intercept after take-off (say, 40 degrees). Using the above method, set the heading on the OBI to 040.

Let's assume you've been cleared to taxi to runway 36. This requires a right turn to the taxiway that runs parallel to runways 4 and 22, another right turn to taxi along that taxiway to the takeoff area on Runway 36, and finally, a sharp right turn to get into takeoff position on the runway. When

Figure 8.2 Using the radar view you can orient yourself on the airport and determine how to get to the active runway. Notice the fueling facility in the upper right-hand corner.

you taxi, be sure to use the radar view (press 4) with a clear picture of the entire airport, and then use the rudder keys (C for left and M for right), pressing them two or three times depending on how tightly you want to turn. Press the right arrow key only once to open the throttle just enough to get the aircraft moving. Once it starts, it's a bit difficult to stop. At a reasonable distance from the point at which you want to stop, use the left arrow key to cut the throttle to idle and use the brake (space bar) repeatedly. The slower you taxi, the easier it is to stay on the taxi way.

If you stray off the taxi way, or if your final take-off position is not exactly on the center line of the runway, don't worry too much about it. Such precision, while important in the real world, is rather difficult to achieve with the simulator. If you find you're having a lot of trouble taxiing, continue to taxi all over the airport until you get the feel for the amount of control input required. Once ready for take-off, advance the throttle gradually and watch the airspeed indicator. When its needle passes through 50 knots, add a little back pressure (B key) to achieve lift-off, and then adjust your elevator trim to achieve a nice and steady 500 fpm rate of climb.

Continue to climb straight ahead until you reach 2,000 feet altitude. Level off by reducing the throttle until the vertical speed indicator remains fairly steady on the zero indicator. Once level flight has been established, initiate a right turn, using approximately 30 degrees of bank. You need to press the H key only once and then watch the turn-and-bank indicator and the actual horizon through the windshield. Press the G key when the appropriate bank angle has been achieved. You may need to add a bit of throttle to prevent the aircraft from losing altitude. Always scan all your primary instruments, especially during turns, to make the necessary corrections before finding yourself in an unusual attitude.

Check your DG (not the magnetic compass) for the amount of heading change. You want to achieve a heading of 90 degrees to intercept the 040-degree radial from the VOR. When the DG indicator comes close to that figure, use the F key to induce opposite aileron to stop the turn. You may overshoot the 90-degree heading. If so, make the appropriate corrections until you've established yourself on it and the DG indicates 090 (and eventually the magnetic compass will settle on 090, too).

Now just fly straight ahead. It will probably take quite a while, but eventually you'll see some movement in the OBI needle as it begins to move toward the center. Start a shallow left turn (F key) to take up your new heading of 40 degrees. If you do it just right, you should be established in straight and level flight on a heading of 40 degrees by the time the OBI needle is centered. If not, make whatever heading corrections are necessary, turning to the right if the needle is to the right of center, or to the left if it's to the left of center.

Once you've established yourself on course, keep flying in that direction, making sure that the airplane is staying within 100 or so feet of the selected altitude. (Are you still at 2,000 feet? You should be.) Whenever you think you're ready to turn back, reset the heading indication on your OBI to 220 and watch the FR indication changing to TO (this portion of the OBI is referred to as the ambiguity meter). At this point the needle will again be centered if, you've not strayed from your course. Make a 180-degree turn to the right or left (most pilots prefer left turns because it's easier to see where one is going and to recognize possible traffic conflicts while sitting in the left seat). Since such a turn covers quite a bit of distance, you'll probably find that the needle is no longer centered by the time you're straight and level on a heading of 220 degrees.

You now have two choices. You can either continue to turn until the needle is once more about to center and then level off to a heading of 220 degrees; or you can slew the heading indicator on the OBI (CTRL V and <

or >) until the needle centers, and then make a heading correction to the heading indication on the OBI which will take you back to the Champaign VOR. When you've established yourself on the appropriate heading and the OBI needle remains centered, simply continue to fly straight ahead and level. It will probably seem to take some time, but eventually you'll see something that looks like the Willard airport in the distance, coming closer and closer at snail's pace. When you're over it you might want to use the radar view to look down at the airport. Watch your OBI. As you pass over the VOR, the ambiguity meter will change from TO to FR, indicating station passage.

Unless you're a glutton for punishment, don't bother to land. The techniques of making approaches to a landing are the subject of another chapter. What we've done up to this point should be enough for this session.

Cross-Country
New York to Boston

Cross country, in aviation terminology, is any flight from one airport to another, even if they're only a few miles apart. Let's take off from New York's LaGuardia Airport and fly to Boston's Logan International. Before we start, let me remind you of the old saying: "Flying represents hours of boredom punctuated by moments of stark terror." As long as we're using a simulator, we can forget about the stark terror, but the boredom part still remains. The simulator operates in real time, which means that flying the 163 nautical miles from New York to Boston at an average ground speed of 125 knots will take an hour and 20 minutes; that's just about exactly what it took me to make that flight in preparation for this chapter.

After activating the program press the ESC key to relocate at LaGuardia Airport, as described in Chapter 4 (17091 north and 21026 east at an elevation of 22 feet). Press the ESC key again, and you'll be at the LaGuardia Airport. Before taking off, set the NAV 1 radio to 108.80, the frequency for the Bridgeport VOR (BDR), the OBI 1 to 063 degrees, representing the approximate magnetic heading toward Bridgeport, the NAV 2 radio to 113.10, the frequency for the LaGuardia VOR (LGA), and the OBI 2 to 063 degrees. While still on the ground, point the nose of the airplane to 040 degrees. We'll assume that runway 4 is the active runway. Take off in the usual manner and climb on a heading of 040 degrees until clear of the airport traffic area. Then make a heading correction to 075 degrees to intercept the 063 radial from LaGuardia (Figure 9.1).

Once you are close to intercepting the radial, make a heading correction to the left and settle down in straight flight on a 063 heading with the CDI (course deviation indicator needle) in the OBI 2 more or less centered.

Figure 9.1 After take-off from LaGuardia, using radar you can look down and see your airplane in relation to the airport.

Depending on your altitude at that point, OBI 1 may still indicate OFF, because at low altitudes you will be beyond the acceptable reception distance. Keep climbing; pretty soon the OBI 1 should come to life, indicating TO with the CDI also more or less centered. If it is not, change the heading reading in the OBI 1 until the needle does center. Then do some very shallow banks to make whatever heading correction is necessary to bring the reading in the DG to coincide with that on the OBI 1.

Once all that has been accomplished, you should be at an acceptable cruising altitude. Level off to a comfortable cruise at something like 65 percent of power, or 2250 rpm. Press the 4 key to enter the radar view, and press the < key a few times to get a good look at your position with reference to the ground and the coast line below (Figure 9.2). Make most of the flight using this radar view; the view straight ahead is simply too dull, with rarely anything meaningful to look at.

It takes about 20 minutes to get to the Bridgeport VOR, located right on the airport (Figure 9.3). If you've steered with a reasonable degree of precision, you should now be right on top of the airport. A few presses on

Figure 9.2 After leaving LaGuardia you will fly over the Long Island Sound toward Bridgeport.

the > key will bring the runway pattern up close and into fairly clear view. After passing the airport, once more use the < key several times to see a greater portion of the ground below.

Reset the NAV 2 radio to 114.90, the frequency for the Hartford VOR (HFD), and then experiment around with the heading display in the OBI 2 until the CDI centers. It should center on an 054 degree heading. Make a shallow bank to the left to establish yourself on that heading, with the DG also reading 054.

It will take another 20 minutes or so to get close to the Hartford VOR. After some 15 minutes of straight and level flight, switch to the 45-degree left view (press 5 R). You'll see something resembling a city in the distance. Revert back to the radar view; that same city will appear to the left of the aircraft symbol. The airport is visible because the VOR is several miles southeast of it.

By the way, if you're planning to make a number of the different cross country flights described here (or any others of your choice), I'd suggest purchasing the appropriate Sectional Charts from the FBO at the nearest

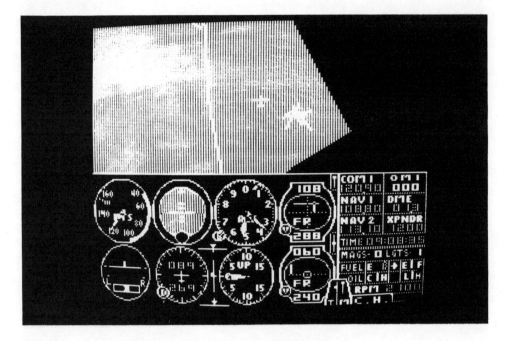

Figure 9.3 Some 20 minutes later you are flying over the Bridgeport Airport.

airport. Sectionals display all the meaningful ground features (such as cities, roads, rivers, etc.) It will make the flight more interesting if you compare these features with the representations in the radar-view mode.

Having once left Hartford behind, change the frequency on the NAV 1 radio to 112.70, representing the Boston VOR (BOS). Once again experiment with the heading in the OBI 2 until the indication is TO and the CDI is centered. It should be approximately 072 degrees. Now make a shallow heading correction to the left until the reading in the DG coincides with that in the OBI 1. Don't be surprised if the display in it reads OFF, once the frequency in that OBI has been changed to 122.70. There are still some 80 nautical miles to Boston; that distance may be too great to receive the signal from that VOR. Wait to adjust the heading in the OBI until the OFF flag has disappeared. Once that has happend and you have adjusted the heading to the correct TO reading, check the distance to the station now displayed in the DME read-out.

You're flying at a ground speed of approximately 125 knots; it is logical that the time to the station is just a few minutes less than half the

Figure 9.4 Finally, after what has seemed like hours, you will see Boston ahead in the distance.

number of nautical miles shown by the DME, at this point, in the neighborhood of half an hour. You will pass several towns along the way, which will help alleviate the boredom. Occasionally switch to the forward view to see if something resembling the big city of Boston is coming into view. Eventually it appears (Figure 9.4).

When you get within 10 to 15 miles of your distination, you can use the radar view to see first the suburbs and the city, and finally Boston Harbor and Logan International Airport (Figure 9.5). Once actually over the airport, you'll want to take a closer look at the runway pattern (press > several times) to determine which runway is which (Figure 9.6).

For all practical purposes, this is the end of your flight. You might pretend to be cleared to land on runway 22R and then descend, fly a normal pattern and attempt a landing. Since actual landings are nearly impossible to accomplish with the simulator, it may be simpler to abandon the flight by turning the computer off.

This flight was made under VFR conditions, using more or less direct routes between fixes (the four VORs used for navigational guidance). Now

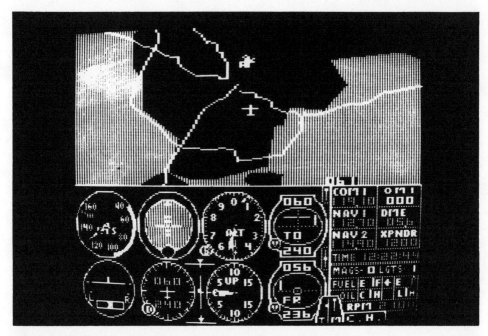

Figure 9.5 Once more using the radar, you approach Boston's Logan Airport.

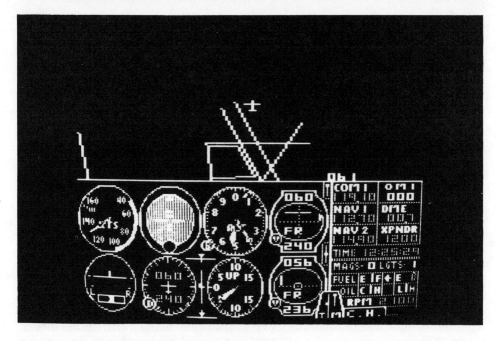

Figure 9.6 As you fly over the airport you can look down at the runway pattern.

check into a motel and get a good night's sleep, because tomorrow you'll fly back to New York. This time the weather will not be quite as good, and the return trip will be made along the airways, flying IFR.

Cross-Country
Boston to New York IFR

Refreshed after a good night's sleep in Boston and a healthy breakfast, return to the airport and make sure the airplane is fully fueled and ready to go. Before leaving the motel call the Boston Flight Service Station (FSS). The weather briefing indicates a solid overcast with bases around 3000 feet and tops reported at about 5000 feet. This indicates that we might be flying in clouds. File an IFR flight plan, asking for Boston direct to Hartford, and from there direct to LaGuardia at an altitude of 6,000 feet. This should keep us above the clouds in happy sunshine.

Back to reality. Start the program, call up the menu (press the ESC key), and locate the plane at Logan International Airport (17899 north and 21853 east at an elevation of 20 feet). Then locate the pointer opposite the CLOUD TOPS 1 category and type 5000. Move the pointer opposite the CLOUD BOTTOMS 1 category and type 3000. Press the ESC key again, and after a few moments hesitation we will be at the Boston airport.

Since this flight is going to be IFR, we can't simply take off and bore holes in the sky. First we must call Clearance Delivery for clearance. To make things more interesting, assume that the clearance will read as follows:

"Piper one two five niner papa is cleared to the LaGuardia Airport via Victor 3 to the BOSOX intersection, Victor 292 to the Putnam VOR, Victor 308 to the Norwich VOR, Victor 16 to the Riverhead VOR and from there direct to LaGuardia. After take-off climb at runway heading to 2,500 feet, turn right to a heading of 300 degrees to intercept the 266-degree radial from Boston VOR. Remain at 2,500 until further clearance. Over."

If this were the real world, you would now be expected to read this clearance back to the controller, who would continue with:

"Clearance correct. Contact tower when ready."

Take some time and look at your flight chart (see Figure 10.1) to see

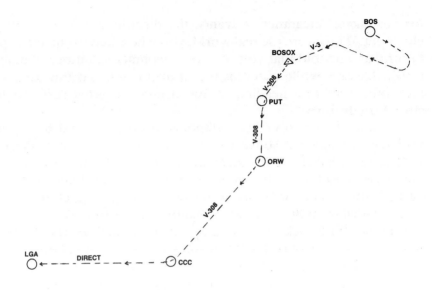

Figure 10.1 The IFR route (not to scale) for the return flight from Boston to New York.

graphically what you're supposed to be doing. By monitoring the tower you know the active runway is runway 15; you're expected to climb to 2,500 feet in a southeasterly direction before turning to intercept the 266-degree radial from BOS. You can also see that the BOSOX intersection is determined by the 142-degree radial from the Gardner VOR (GDM) or the 57-degree radial from the Putnam VOR (PUT). Since the latter is the one you'll use later, use that radial to determine the location of the intersection.

With all this more or less memorized, set the NAV 1 radio to the BOS VOR frequency (112.70). It controls the DME, which will tell you the plane's distance from that VOR in nautical miles. This is helpful because the chart shows the distance from BOS to BOSOX to be 24 nautical miles. Set the heading in the OBI 1 to 266, the radial you're expected to intercept. Change the frequency in NAV 2 to the PUT VOR (117.40), and the heading in the OBI 2 to 057. While OBI 1 will come to life immediately, you won't be able to pick up PUT on the OBI 2 until you've reached a reasonable altitude.

Now call the ground control for clearance to taxi to runway 15. Using the radar view, locate yourself on the airport and taxi to the northwest end of the active runway. After the usual preflight checks, contact the tower for our takeoff clearance. Advance the throttle and perform the usual climb-out. If this were the real world, you'd be asked to conact Departure Control and be told that you're in radar contact. Continue climbing at a reasonable rate, while watching the altimeter. When it approaches 2,500 feet, turn right to a heading of 300 degrees, making sure not to lose altitude in the turn.

Assume that you've been climbing at 80 knots and 600 fpm. You should have reached 2,500 feet a little over four minutes, meaning that the DME should read about 5.6 nm from the BOS VOR. Once established on a 300-degree heading, continue in straight and level flight, watching the CDI in OBI 1. When it begins to move, you're getting close to the 266-degree radial from BOS. Initiate a shallow left turn so as not to overshoot that radial. With luck you'll end up on that radial in straight and level flight. Departure Control will call and tell you to contact Boston Center, now on 121.35. Comply; Boston Center replies with: "Piper one two five niner, radar contact. Climb to and maintain 4,000 now. Over."

Shove the throttle all the way in and start to climb to 4,000 feet. Keep an eagle eye out for the CDI in OBI 2, which is likely to come to life before you've reached the new altitude, which places you right smack in the middle of the cloud layer. All views from the aircraft, other than the radar view, are totally obscured by clouds.

The CDI on the OBI 2 starts to move just as you reach the new altitude and are leveling out. You're approaching BOSOX; turn left to a heading of 237 degrees (the reciprocal of 057), then switch the heading in the OBI 2 to 237 to fly the TO indication. There would be nothing wrong with choosing to leave it on 057 and flying the FR indication. The distance from BOSOX to PUT is 22 nm; it'll take about 10 minutes to get there.

If this were a real flight, you'd want to call Boston Center (or possibly New York Center if you've already been turned over to it) and ask for a higher altitude to get out of the clouds. Depending on conflicting traffic you might get it. In the simulation, the convenience of the radar view gives a clear picture of the ground below.

On the way to the PUT VOR, we'll reset the NAV 1 radio to the frequency for the next fix, the Norwich (ORW) VOR (110.00), and the heading on OBI 1 to 210 degrees. Once you reach PUT, make a slight heading correction to the left and establish yourselves on the 210-degree bearing to ORW. The distance from PUT to ORW is 15 nm, and the time en route is about seven minutes. During this time reset NAV 2 to the frequency for the Calverton (CCC) VOR (117.20), and the heading in OBI 2 to 241 degrees in order to be prepared to find the next fix. The distance between ORW and CCC is 57 nm. Depending on the quality of the navigation radio, you may not be able to pick up CCC from that distance. To be sure you head in the right direction having passed ORW, set the heading OBI 1 also to 241 degrees and then fly that radial outbound until CCC comes in.

So far we've assumed that there is no wind. If you were fighting a 10- or 15-knot crosswind, you'd find the aircraft constantly drifting off course. You'd have to make appropriate heading corrections to stay on the VOR radial or bearing. Such heading corrections would result in the fact that the heading readings in the OBIs and the DG would differ. We'll take a closer look at that phenomenon during other cross countries.

While flying between Norwich and Calverton you might want to dial the frequency for CCC into NAV 1 to monitor the remaining distance to the station on the DME read-out. This will give you a good idea of the time remaining on this leg before having to make final course corrections for the direct leg to LGA. Once approaching Calverton (after some 22 minutes) reset NAV 2 to the LGA VOR (113.10) and the heading in OBI 2 to 285 degrees, the approximate heading toward LGA. When station passage is imminent, make a course correction to the right until the needle in OBI 2 begins to center, then roll out on that heading. Since NAV 2 does not control the DME, you won't be reading the remaining distance to our destination. Since the distance from CCC to LGA can be measured on the chart (resulting in 43 nm), you know that when the DME reads 43.0 you'll be just about over the destination airport.

At this point you'd have to contact LaGuardia Approach Control and follow its instructions to land. However, we'll simply decide that you have arrived. There is one thing we neglected yesterday which should always be done: once near the destination airport, we should always switch to the fullest tank. It can be disconcerting to run a tank dry on final approach (I speak from experience).

We'll make more cross-country flights, some with wind and turbulence, and some including approaches to an actual landing.

A Look at Approach Techniques

So far we've carefully managed to avoid the complicated procedure of flying approaches to a landing. It is difficult to get a clear idea of the turning radius of the aircraft at different degrees of bank, a problem of overriding importance in flying approaches, whether IFR or VFR. Lets look at VFR approaches first.

When operating under visual flight rules, there are two basic types of approaches commonly used. One involves flying a left or right pattern, and a other is the straight-in approach. Before getting ready to attempt an approach, we must know the airport elevation; we should plan to flare and pretend to land somewhere between 25 and 50 feet above the ground, because in this simulator it is impossible to gauge the exact distance between the ground and the wheels of the aircraft. Without the ability to accurately judge that distance, a safe landing is impossible.

Preparatory to starting any VFR approach establish yourself in straight and level flight at approximately 800 to 1,000 feet above the airport elevation. Slow the aircraft down to somewhere around 80 knots. Here are some of the initial maneuvers to practice:

1. Descend from your cruising altitude to pattern altitude, level off and reduce speed to 80 knots without losing additional altitude. During the descent pull the throttle nearly all the way back, and then use the elevator to set up a smooth glide at about 500 fpm and somewhere around your previous cruising speed of 125 knots. When within about 100 feet of your pattern altitude, use elevator control to level off and add sufficient throttle to keep the aircraft in level flight at the reduced speed. You'll find this is not easy, and I suggest that you practice this maneuver a couple dozen times, until the appropriate throttle settings and elevator positions are second nature.

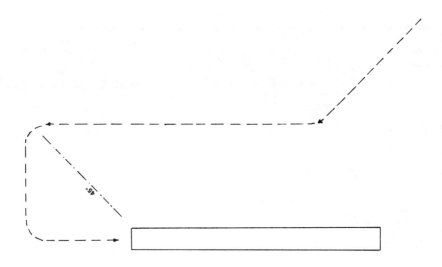

Figure 11.1 The standard VFR airport traffic pattern. When the approach end of the runway is at approximately 45 degrees behind, the turn to base leg is initiated.

2. Next practice flying a pattern. Since the landing patterns at most airports require left turns, stick to left-turn patterns. Use the radar view of some airport or other fix on the ground and adjust it to a size that seems to best represent about 1,000 feet of altitude. Look at Figure 11.1 and try to fly that pattern. At first your turns will likely be too shallow or too tight and you'll overshoot or under-shoot the desired position for the next pattern leg. Don't be sloppy; a good pattern is entered at a 45-degree angle to the downwind leg. That leg is flown until the approach end of the runway is approximately 45 degrees behind your left shoulder. Make a 90-degree left turn to the base leg, and continue straight ahead until the runway is just ahead and to your left. Then make another 90-degree turn to the left to establish yourself on the final approach. For the time being, don't bother to descend. Simply practice flying this pattern until it, too, becomes easy.

3. Once you're satisfied with your mastery of the first two maneu-vers, add vertical control to the pattern. The downwind leg should be flown straight and level, and the descent from pattern altitude should not be started until you're established on your base leg. Once on base, reduce throttle a few notches to cause the aircraft to gradually descend, but not too fast. When making the turn to the

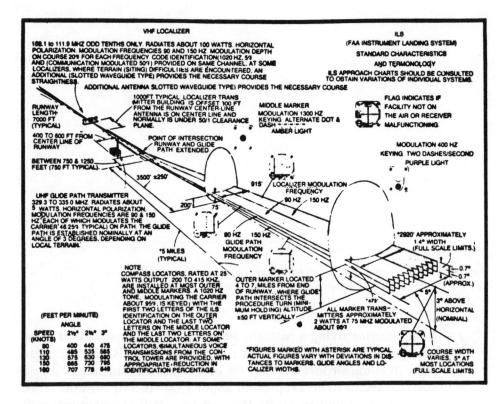

Figure 11.2 A diagram showing the ILS approach system. (Courtesy FAA).

final approach, you may have to add a bit of back pressure (up elevator) to prevent an excessive loss of altitude in the turn. Once established on final, you'll want to reduce throttle all the way or nearly all the way, using the elevator as a speed control to set up a smooth glide at about 60 knots. In the real world, this would permit you to flare a few inches above the runway to a full stall landing.

Landings are the most difficult phase of any flight and should be practiced many times. Don't bother with straight-in approaches on the simulator; the forward view from the cockpit does not permit a clear enough idea of the distance to the runway to safely fly a straight-in approach.

Instrument approaches require us to locate ourselves at given positions relative to ground-based navigation aids, and at predetermined altitudes, while having no visual contact with the ground and the airport. There are precision instrument approaches and non-precision instrument approaches. Precision approaches are those that make use of ground-based altitude control—a glide slope. (There are also precision radar approaches, but they cannot be flown on this simulator.) A *glide slope* is an electronic beam that produces a display in the cockpit. On our simulator, this takes the form of a horizontal bar in OBI 1. It is used in conjunction with the localizer, which provides lateral guidance. The two together constitute a full ILS (Instrument Landing System). (See Figure 11.2.) The minimum altitude to which an aircraft may descend during a precision approach without obtaining visual contact with the airport is referred to as *Decision Height* (DH), and it varies among different airports.

A non-precision approach is one in which the ground-based navigation aids provide only lateral guidance. It is up to the pilot to be constantly aware of the altitude at which he or she is performing the maneuvers that constitute the prescribed approach procedure. The minimum altitude to which an aircraft may descend during a non-precision approach with no visual contact is called *Minimum Descent Altitude* (MDA). The MDA is always higher than the DH.

Regardless of the type of approach, the aircraft at DH or MDA may continue in level flight toward the airport until reaching the Missed Approach Point (MAP). If at that point visual contact has not been established, the aircraft must execute a missed approach, following certain predetermined maneuvers.

There are literally thousands of different instrument approach procedures, and for each there is something referred to as an *approach plate*, a small chart that graphically depicts the maneuvers, laterally and vertically, that constitute the approach procedure. All we can really do on our simulator is practice a few typical routines to get a feel for what it's like. Since instrument approaches are normally flown under actual or simulated instrument conditions, the view out the cockpit is superfluous and, in fact, distracting. It might be a good idea to select a low overcast from the menu in order to cause the upper portion of the display to produce no images.

Figures 11.3 through 11.12 depict several typical instrument approaches. Before attempting to fly any one of them, study them in great detail. Once in the air it's too late to try to figure out what you're supposed to be doing. Since this is a simulation, you can press the P key at any time to stop everything and decide what it is you're supposed to be doing next, but in a real airplane you wouldn't have that convenience. It is bad practice to make use of the pause feature.

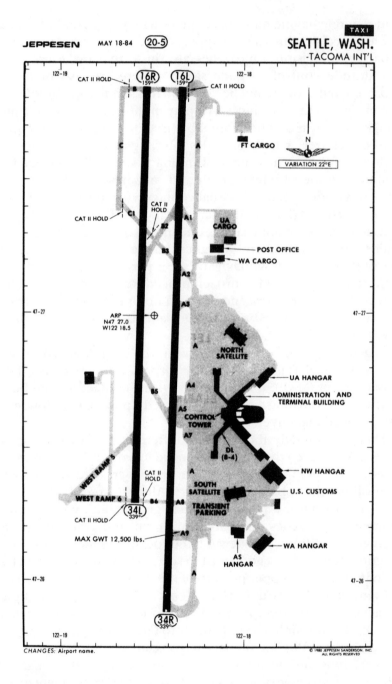

Figure 11.3 A detailed representation of the Seattle-Tacoma International Airport. (Copyright© Jeppesen & Co.).

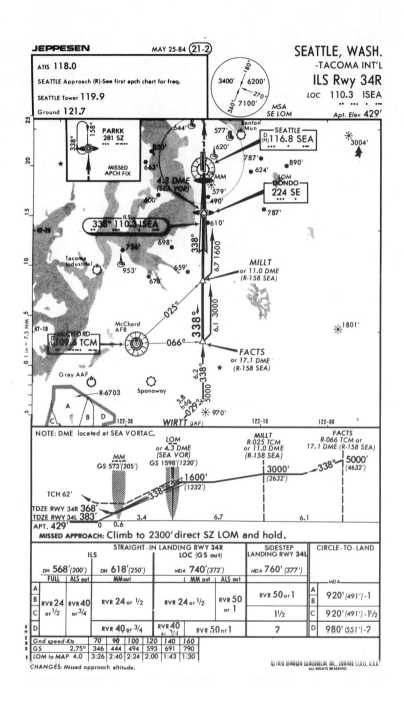

Figure 11.4 The Seattle-Tacoma ILS Runway 34 approach chart. (Copyright© 1985 Jeppeson & Co.).

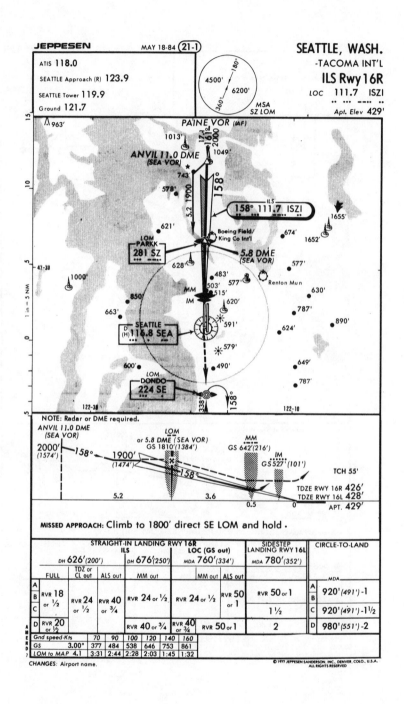

Figure 11.5 The Seattle-Tacoma ILS Runway 16R approach chart. (Copyright©1985 Jeppeson & Co.).

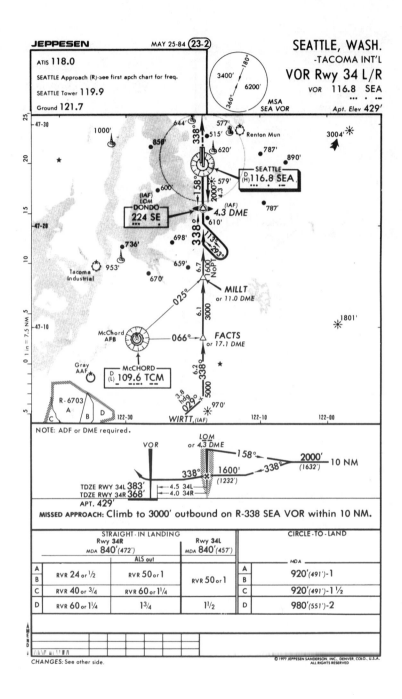

Figure 11.6 The Seattle-Tacoma VOR Runway 34 approach chart. (Copyright©1985 Jeppeson & Co.).

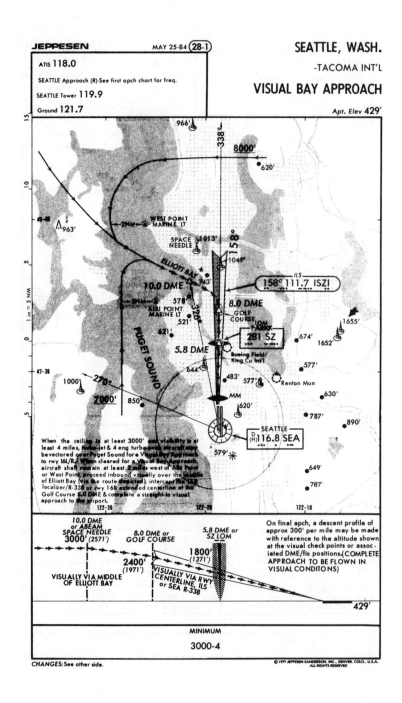

Figure 11.7 The Seattle-Tacoma visual bay approach chart. (Copyright©1985 Jeppeson & Co.).

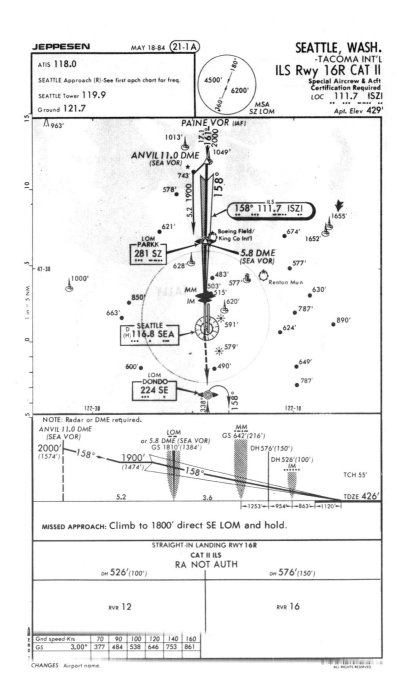

Figure 11.8 The Seattle-Tacoma ILS Runway 16R CAT II approach chart. (Copyright©1985 Jeppeson & Co.).

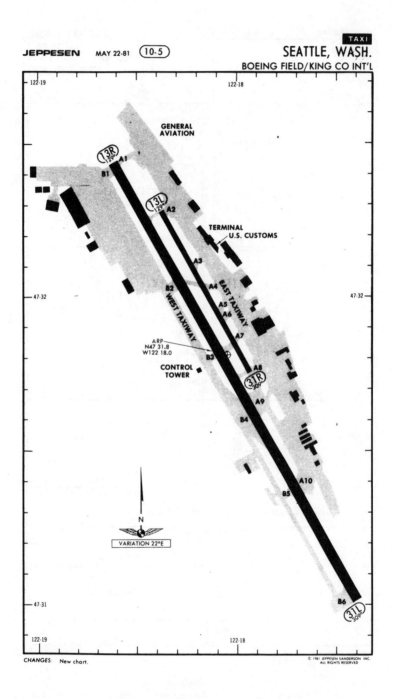

Figure 11.9 The layout of Seattle's Boeing Field. (Copyright©1985 Jeppeson & Co.).

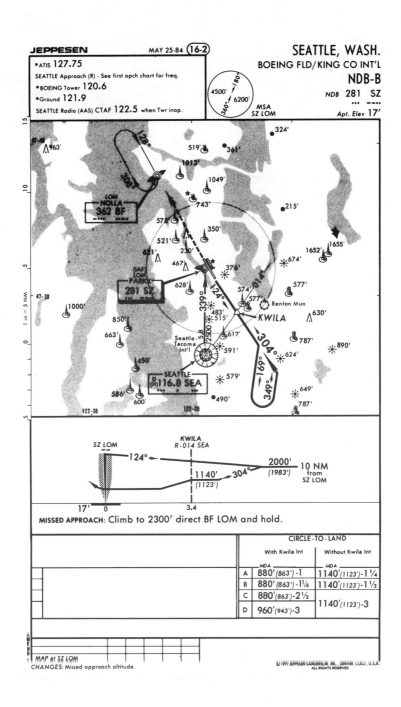

Figure 11.10 The Boeing Field NDB-B approach chart. (Copyright©1985 Jeppeson & Co.).

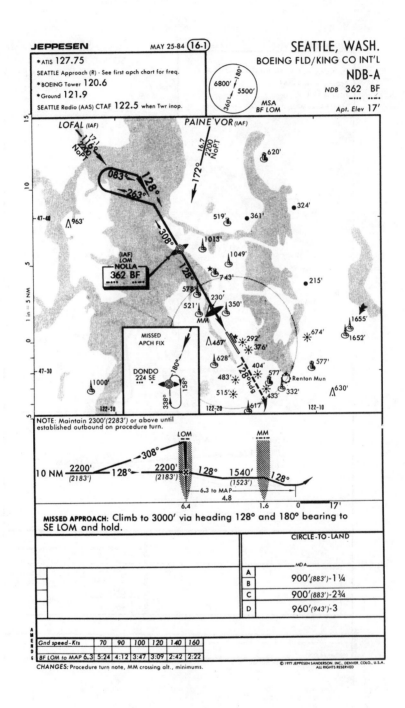

Figure 11.11 The Boeing Field NDB-A approach chart. (Copyright©1985 Jeppeson & Co.).

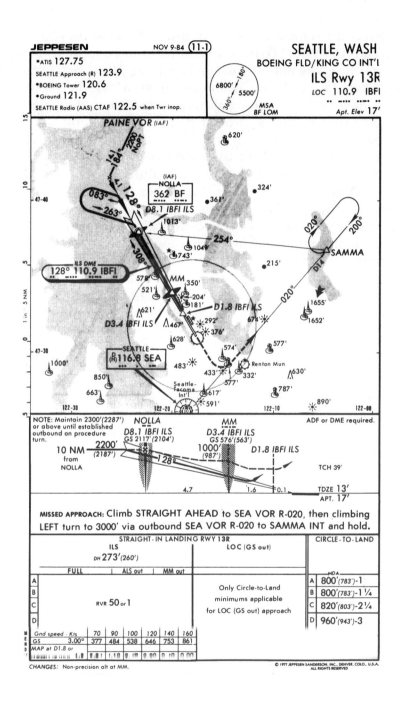

Figure 11.12 The Boeing Field ILS Runway 13R approach chart. (Copyright©1985 Jeppeson & Co.).

Every instrument approach begins by locating the plane at a specific fix at a given altitude. To fly any of the instrument approaches available on the simulator, you must, by using the editing menu, first locate the plane somewhere in the vicinity of the airport to which the approach is to be flown, preferably at cruising altitude, which should be higher than the altitude listed on the approach plate in association with the Initial Approach Fix (IAF). Then, depending on the type of approach, you'll use your navigation radio and OBI or the ADF to fly to the IAF while, at the same time, descending to the published altitude. Once over the fix at the correct altitude, take up the published headings and follow the instructions on the approach plate. This requires an intimate understanding of how to precisely control your aircraft horizontally and vertically. Things tend to happen rather quickly during approach.

In the real world when you are approaching a major airport with an operating Approach Control, you would usually be in radar contact, and would be told when to turn to what heading, and when to descend to what altitude. These instructions frequently differ from those shown on the charts. The radar controller knows the position and altitude of each aircraft in the area, and approaches can often be speeded up by eliminating all or some of the published procedures.

The *Flight Physics and Aircraft Control* booklet that is part of the *Flight Simulator II* package includes detailed descriptions of two instrument approaches—a VOR/DME approach and an ILS approach to the airport at Champaign, Illinois. I suggest that you practice those and then try some of the others for which approach plates are reproduced here.

Whether or not you want to spend a great deal of time practicing instrument approaches depends on whether you are or intend to become an instrument-rated pilot. Much of whatever degree of proficiency you may attain cannot readily be transferred to a real airplane or even a real simulator. On the other hand, you will gain an insight into the intricacies of instrument approaches, and you'll learn not to be intimidated by approach plates that, at first glance, seem to be incomprehensible.

From Van Nuys to San Diego with Wind and ADF

Now we'll fly from Van Nuys to San Diego in a fairly strong crosswind, and along part of the route we'll navigate with the automatic direction finder (ADF). First locate the plane at the departure airport in Van Nuys, California. From the menu, key in 15498 north and 5811 east at an elevation of 799 feet. On the next page of the menu, key in the wind direction and velocity data: wind 3 blowing at 35 knots from 180 degrees above 4,000 feet, WIND 2 blowing at 30 knots from 190 degrees above 3,000 feet. WIND 1 blowing at 30 knots from 195 degrees above 1,500 feet, and the SURFACE WIND blowing at 10 knots from 160 degrees.

Next move the pointer to the ADF category, and press 1 to replace NAV 2 and its associated OBI with the ADF and its display. Then press the ESC key to return to the program. After a few moments you're sitting on the ground at Van Nuys.

Take a look at the instrument panel; the ADF and its readout have displaced the second nav radio and OBI. Press 4 and the < keys a number of times; your location with reference to the runways will appear. With runway 16 the active runway, you'll have to taxi to the take off end of that runway, but for the time being there are other preparations to make. First enter the frequency for the Escondido (EKG) non-directional beacon (NDB), which is 374. Press CTRL A once to adjust the first digit (hundreds) to 3. Then press CTRL A twice to adjust the second digit (tens) to 7, and finally press CTRL A three times to adjust the third digit (ones) to 4. Note that the number displayed under ADF in the display remains unchanged, because you're much too far from the NDB to pick up its signal.

Next tune NAV 1 to the LAX VOR frequency (113.60) and change the heading in the OBI to 160 degrees, the approximate heading toward Los Angeles International Airport. Since it is only 17.3 miles from Van Nuys, as shown on the DME, the OBI comes to life immediately and the CDI will

Figure 12.1 The second OBI has been replaced by an ADF display. The airplane is over the Marina, having flown over Santa Monica Airport. Hughes single runway airport is to the left of the airplane, with Los Angeles and Torrance airports and to front left.

more or less center. With the radios all set for the time being, key in one notch of throttle and taxi around the airport area until you're positioned on the take off end of runway 16R or 16L. Be sure the heading indication on the DG reads 160 with the runway ahead of the aircraft. Take off in the usual manner, continuing on the runway heading and climbing to an altitude between 2,500 and 5,000 feet. You're going to fly through what is known as the VFR corridor that goes right over the center of Los Angeles International Airport between 2,500 and 5,000 feet.

After five minutes or so, the single runway Santa Monica airport comes into view. Fly right over it; by manipulating the radar view you can look at its runway and taxiway (Figure 12.1). A few minutes later you'll pass the Marina and Hughes Aircraft's private airport (Figure 12.1). After only another minute or two the huge Los Angeles airport appears and you may have to make a minor heading correction to fly right over the middle of it (Figure 12.2).

So far there has been no noticable wind effect because the wind has been more or less on your nose, thus slowing the plane down but not

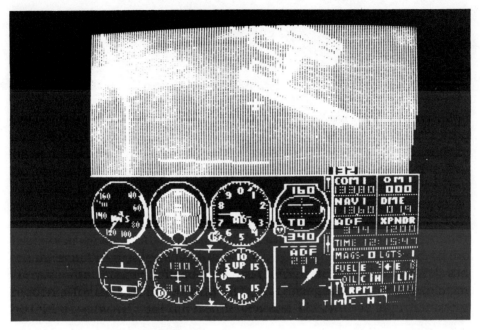

Figure 12.2 After several small heading corrections the airplane is headed straight for the VFR corridor overlying Los Angeles International Airport.

affecting your course. Once safely past LAX tune NAV 1 to 115.70, the frequency for the Seal Beach VOR (SLI), and change the heading display in the OBI downward until the needle centers. This will be approximately 096 degrees, depending on how far past LAX you are at this point. Now take up this new heading and sit back and relax. Probably just before reaching SLI VOR you'll suddenly notice that the indication of the ADF display has changed; the needle will be somewhat to the right of center and the number reading something like 009. That means the nose of the aircraft must be turned nine degrees to the right to point straight toward the NDB. (If that display were showing 347, for instance, it would indicate 13 degrees to the left, the difference between 360 and 347).

Make a heading correction to the right until the ADF needle is centered (pointing straight up) and the display reads 000. Theoretically, you could now navigate by the ADF alone and ignore the nav radio altogether. That's not a good idea, because ADF readings have a habit of occasionally becoming erratic, especially if there's a thunderstorm somewhere in the area. As a back up, tune NAV 1 to the frequency for the Oceanside VOR

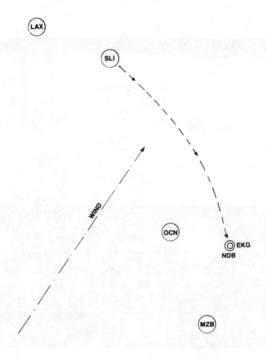

Figure 12.3 When navigating by ADF, a crosswind tends to result in a curved flight path.

(OCN),115.30, and manipulate the heading in the OBI until the needle centers. Another advantage of using the nav radio with the ADF is that the DME gives at least an approximation of the plane's progress. I say an approximation, because as long as you continue to use the ADF as the primary navigation instrument, you won't be flying directly toward the VOR.

After a while you'll notice a new phenomenon. Despite the fact that you've been continuing to fly on the same heading, approximately 120 degrees, the ADF needle insists on drifting to the right, showing several degrees of deflection, requiring repeated heading corrections to the right. The reason is the right crosswind which keeps blowing the plane to the left of the course (see Figure 12.3). This illustrates the difference between navigation by VOR and ADF. If you were using the bearing to a VOR as the primary navigation aid, you'd be keeping the CDI centered, flying a straight line rather than a shallow curve. You'd find that the actual heading, as displayed by the DG, would be several degrees greater than that shown on the OBI, because the wind causes you to actually fly slightly sideways (see Figure 12.4).

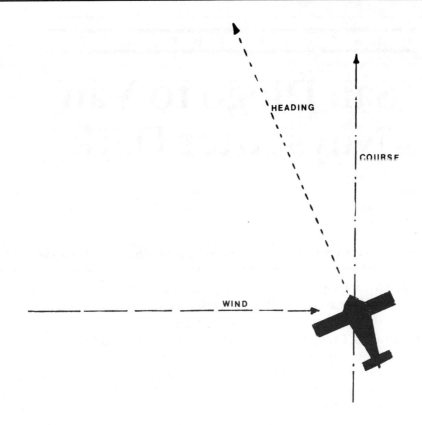

Figure 12.4 Crosswinds cause the aircraft to angle into the wind in order to stay on course. "Heading" is the direction in which the nose of the airplane is pointing, and "course" is the flight path over the ground.

Once you approach the OCN VOR, you can switch NAV 1 to 117.80, the frequency for the Mission Bay VOR (MZB), located just southwest of San Diego's Lindbergh International Airport. At this point you might as well abandon the ADF (just ignore it) and fly a direct route toward MZB, using the OBI indication for navigational guidance.

The straight-line distance from Van Nuys to San Diego is 110 nm which, at an average ground speed of 125 knots, would take 53 minutes. But you did not fly a straight line, adding up to six or seven miles to the total distance. In addition, you've been fighting a quartering headwind that probably reduced the actual ground speed by some 20 knots. This means that by the time you arrive over San Diego, you've flown approximately 117 nm at a ground speed of 105 knots. Thus, the total trip will have taken approximately one hour and seven minutes.

In order to add a new experience, we'll make a return trip after dark.

San Diego to Van Nuys After Dark

To make a return flight after dark, taking off from San Diego at 10 PM and arriving in Van Nuys about an hour later, we'll assume the winds have calmed down, as they usually do at night. With the menu displayed first locate the plane at San Diego (14761 north and 6102 east at an elevation of 15 feet) and then, on the second page, set the time at 2200 hours (10 PM) and zero minutes. Pressing the ESC key places the plane at the airport with only rows of lights appearing in front. Call up the radar view; the runway lights come into view, making it easy to determine the direction in which to taxi to get to the takeoff end of runway 27 (nearly always the active runway at Lindbergh).

Since we have not chosen to use the ADF on this flight, set NAV 1 to 115.30, the OCN VOR and NAV 2 to 117.80, the MZB VOR, adjusting the heading in OBI 1 to 329, which is the heading from MZB to OCN. Set OBI 2 to the heading that produces a centered CDE with the TO indication; that will be the heading to fly after lift-off to pass over the MZB VOR. After lift-off make the slight heading correction that will take you to the MZB VOR, and then establish the plane on the 329-degree radial from MZB. It's important to make sure you don't drift too far to the right of that radial; there are mountains nearby as well as the Camp Pendleton restricted area.

During the rest of the flight, toggle back and forth between the 4 and 5 keys to alternately see the forward and the radar views that show highways, identified by the headlights of cars and the lighted airports along the way (Figure 13.1). Once level at cruising altitude, tune NAV 2 to 113.60, the LAX OR frequency, which you'll use after passing over Oceanside. At that point tune NAV 1 also to LAX to receive the DME read-out, and tune NAV 2 to 113.10, the VNY VOR frequency which you'll use after having flown through the VFR corridor over Los Angeles International.

By using the < key repeatedly in the radar view mode, you'll soon be

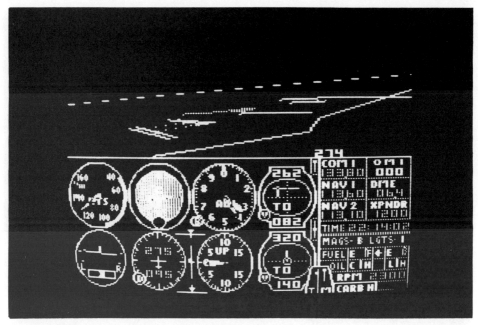

Figure 13.1 Approaching Los Angeles on the night flight from San Diego.

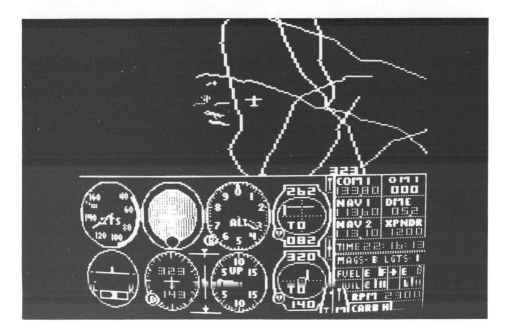

Figure 13.2 The brightly-lit Los Angeles freeway system.

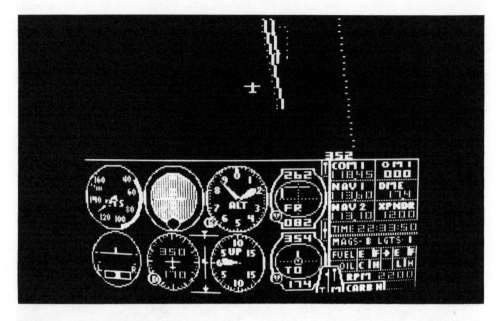

Figure 13.3 Downwind at Van Nuys.

able to see the entire Los Angeles freeway system with LAX and other airports as small lighted blobs. Within a few miles of LAX, use the > key to get a closer look at its runway pattern at night. Once past that airport make a heading correction to about 340 degrees, and after flying over the lighted runways of Hughes and Santa Monica airports, you'll be heading straight toward Van Nuys (Figure 13.2).

Once approaching the airport, attempt to fly a downward leg to the left of the runway, then make a 180-degree turn to the right and try to get lined up with the runway for landing (Figure 13.3). If you don't succeed the first time, do it again and, if necessary, several times more until you get the hang of it, gauging the degree of bank necessary to fly a decent pattern. Flying a good pattern is half the battle in making a good landing (even though actual landings can not readily be accomplished on this simulator). Practicing flying landing patterns with this simulator is easier in the night mode, because the runway patterns are more clearly defined than they are in the day mode.

The Angle of Attack

While this book is certainly not intended to teach you the finer points of how to fly an airplane, there's one subject that, for the benefit of non-pilot readers, should be discussed in some detail. That subject is known as the *angle of attack*. Without careful attention to it, you're going to crash your airplane with depressing regularity. Simply stated, the angle of attack is the angle at which the air, or relative wind, strikes the wing. But there's more to it.

Instinctively we tend to feel that when the nose of the airplane is pointed in a given direction, that is the direction in which we're flying. Therefore, the relative wind comes from the direction toward which the nose is pointing. Not so. Whenever we're operating at lower than normal airspeeds, the nose of the airplane will point several degrees above the direction in which we're actually moving through the air. Thus, the angle of attack will be several degrees greater than it is during level cruise.

Another mistaken idea is that some relation exists between the angle of attack and the angle of the nose of the airplane relative to the horizon. No such relation exists. For instance, when zooming upward at high speeds, the nose of the airplane may be pointed quite high; as long as adequate speed is maintained the angle of attack will even so be quite small. Alternatively, the nose of the aircraft may actually be below the horizon, but the angle of attack could be great if airspeed is low (see Figures 14.1 through 14.3).

At high airspeeds, then, the force of the relative wind is considerable, requiring a minimum angle of attack in order to produce the necessary lift. When airspeed is lowered, the force of the relative wind is lessened, requiring ever greater angles of attack in order to produce sufficient lift to keep the airplane flying. But there's a limit: an increase in the angle of attack results in an increase in lift *only* until the angle becomes so great

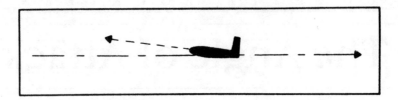

Figure 14.1 When operating at lower than normal airspeed, the nose of the airplane points up, increasing the angle of attack.

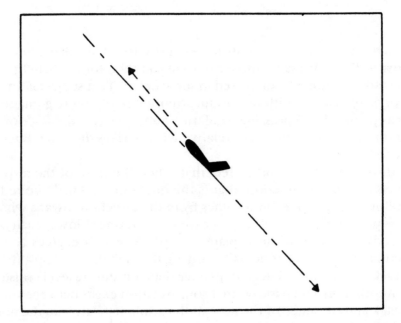

Figure 14.2 In a zoom, the nose of the airplane may point steeply upward, but as long as adequate speed is maintained, the angle of attack remains small.

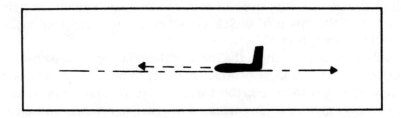

Figure 14.3 At high airspeeds and direction of the nose of the aircraft and the relative wind are virtually parallel, resulting in very small angles of attack.

that the air flowing across the wing becomes disturbed. Severe disturbance of the airflow over the wing causes a sudden and complete loss of lift, resulting in what is known as a stall.

Many real airplanes are equipped with an angle-of-attack indicator. Our simulated airplane is not. In real airplanes lacking an indicator the pilot must rely on the "feel" of the airplane. As long as the angle of attack is sufficiently small to produce adequate lift, the airplane behaves normally, responding to the controls the way it should. Increases in the angle of attack result in diminished control response, telling the pilot to lower the nose of the aircraft to increase airspeed and in turn, decrease the angle of attack.

All this is of little use when we fly our simulator, because the simulated airplane produces no "feel." All we can rely on are the airspeed indicator and the vertical speed indicator. As long as the airspeed is kept high enough, the airplane will want to fly (except, of course, in an inadvertant power drive right into the ground).

When we first start to fly the simulator we tend to want to overcontrol (hit the appropriate keys too many times) because there is always a noticeable delay in the response to any control input. That is a realistic representation of the aircraft being simulated by the program. There are other types of aircraft that produce much faster control responses. The trick is to do things at a reasonable pace. For instance, if we get impatient on take-off, wanting the airplane to get off the ground in a hurry, we may add an excessive amount of back pressure (up elevator—B key), only to find that suddenly the VSI needle climbs all the way to 1,500 fpm, which would indicate too steep a climb. This results in a rapid loss of airspeed and, in turn, an equally rapid increase in the angle of attack. Unless we take corrective action by reducing back pressure (down elevator—T key), the airplane will stall, dropping the nose sharply and, if we're still fairly close to the ground, there won't be enough room to recover and we'll crash.

Controlling the angle of attack means controlling the airspeed. Before attempting long flights, practice airspeed control in climb, level flight, and descent. The throttle is not the primary airspeed control. We need a lot of throttle to get going, but once aloft and in cruise, it is the elevator that controls airspeed and the throttle that controls rates of climb or descent.

Try it: level out at a safe altitude and establish yourself in level cruise at 120 knots, which will probably require approximately 1,950 or 2,000 rpm and possibly a single notch up elevator or elevator trim. Now increase the throttle by pressing the right arrow several times. The result will not be an increase in the airspeed. The aircraft will start to climb, maintaining the current airspeed for a while, but there may actually be a decrease in airspeed if the climb angle starts to steepen.

Now get back to straight and level flight and add a few notches of back pressure. For a moment the airplane will climb a few feet. But very soon it will stop the climb and either fly straight ahead in a nose-high altitude at reduced airspeed, or it may even start to descend.

Practicing this kind of throttle and elevator control is very important, because it teaches us the behavior of real airplanes quite realistically. Practice slow flight at 70 knots without gaining or losing altitude, and then practice controlled descents at that speed. Notice how the nose of the aircraft points high into the sky. But don't overdo it; remember that the simulated aircraft will stall at varying speeds around 50 knots, depending on the flap setting. This is another procedure you should practice. Add flaps when in slow flight, and observe the effect. Depending on the amount of flaps being used, they add both lift and drag, permitting lower airspeeds. Except during severe crosswind conditions, flaps should always be used during approaches to a landing.

No matter what your final goal in the simulator or in a real airplane, being aware of the aircraft's attitude and on top of the flight controls at all times, takes preference over navigation. Even during instrument approaches that frequently require a lot of radio tuning and other distracting activities, flying the airplane comes first. If you lose control of the airplane, all that precision navigation won't do you a bit of good.

Chicago to Seattle in a Hurry

So far we've limited our flights to the navigation areas shown on the four charts supplied with the program. This time, let's be more ambitious. We're parked at Meigs Field in Chicago and have decided to fly to Seattle, Washington (Figure 15.1), a straight-line distance of 1,495 nautical miles. In a real airplane, at an average cruising speed of 125 knots, that would take 11 hours and 58 minutes. At an average fuel flow of 8 gph, we'd burn 95.68 gallons of fuel. Our airplane has a fuel capacity of only 48 gallons, and our maximum range, with 45-minute reserve, is 670 nm. We'll have to stop and refuel at least twice somewhere along the line, not to mention that we'd probably want to get a bite to eat, too.

Checking our route on a low-altitude flight-planning chart, it seems that we'll be flying via Rockford, Sioux Falls, Pierre, Dupree, Miles City, Lewiston, Great Falls and Spokane to Seattle. If I were making that flight I'd want to stop in Pierre and either Lewiston or Great Falls; but that's the real world. The navigation areas available on our simulator extend only a few miles west of Chicago and equally few miles east of Seattle; everything in- between is ''no man's land.''

The simulator *does* include an option that permits flying across these uncharted stretches of the country in a great hurry. That option is the SLEW option on the editing menu.

To start we'll select the regular flight mode, beginning on the runway at Meigs Field. We'll take off toward the north, as usual. Once at a respectable altitude, set the NAV 1 radio to 113.90 and adjust the heading in the OBI 1 until the CDI centers. Make a course correction to the heading indicated by the OBI, and fly toward Chicago's O'Hare airport. The magnetic course from O'Hare to Rockford is 280 degrees. Once over the airport (Figure 15.2), adjust the heading in the OBI to 280 and fly outbound on that radial until you're beyond reception distance and the OFF flag

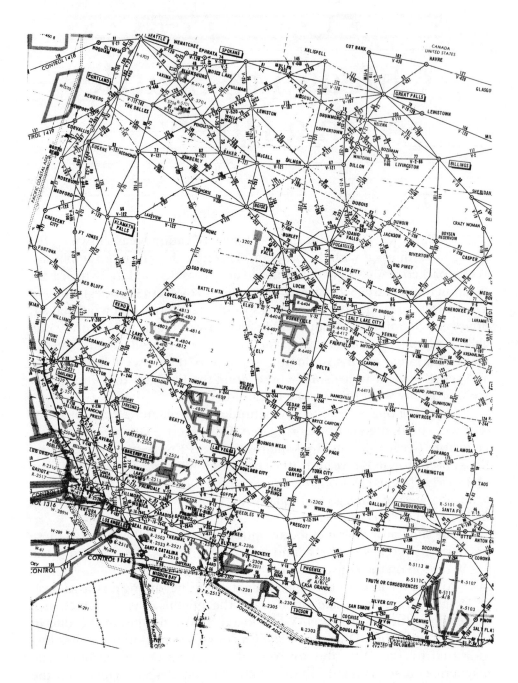

Figure 15.1 A section of the Low-Altitude Planning chart, with Chicago at the far right center and Seattle in the upper lef corner. (Copyright©1985 Jeppeson & Co.).

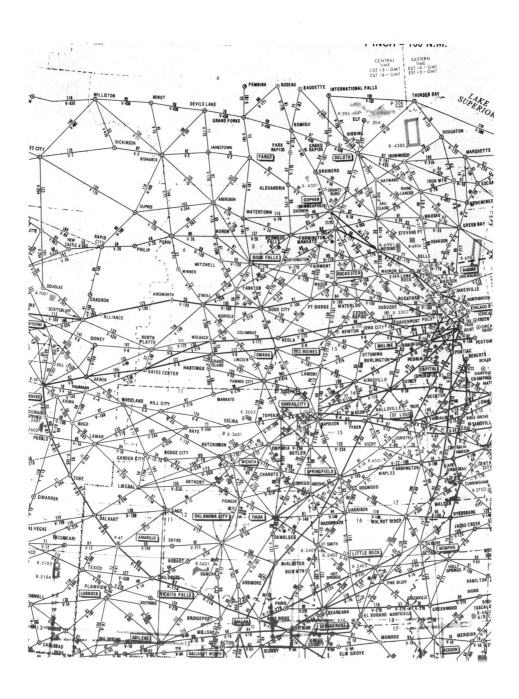

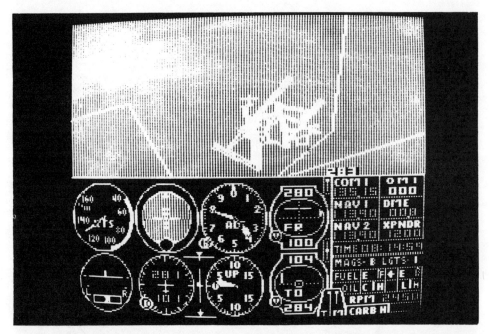

Figure 15.2 Having just flown over O'Hare Airport, the airplane heads west, soon to leave the Chicago navigation area.

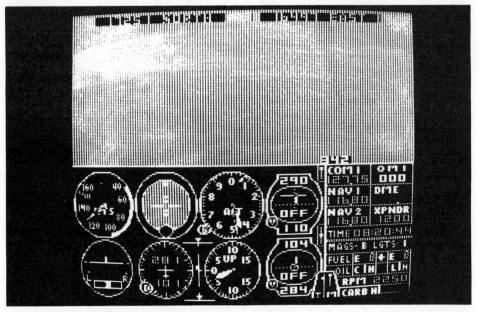

Figure 15.3 In the "slew" mode, the coordinates of the aircraft's present position are displayed at the top of the screen.

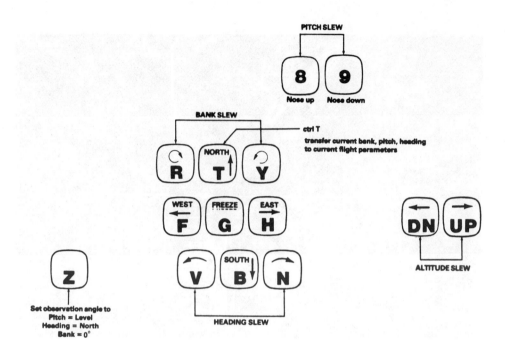

Figure 15.4 Here are the keys used in a slew mode. (Courtesy SubLogic).

appears in the OBI. Depending on your cruising altitude, this should be some 50 nm west of Chicago, which puts the plane within 10 nm of Rockford, Illinois. Even though Rockford has a VOR, it is not included in the data base of our simulator, so you're on your own.

Before doing anything else, look up the VOR frequency for Seattle and tune NAV 1 to 116.80. Nothing will happen; the view out the airplane will also be a big nothing. Now press the ESC key to call up the editing menu. Place the pointer opposite SLEW and type 1 to get into the slewing mode. Now press the ESC key again to return to the plane; at the top of the windshield you will see two data representing the north and east coordinates of your current position, which will most likely be in the 17000 north and 16000 east range (Figure 15.3). Now check the coordinates for Seattle; they are 21343 north and 6584 east, which is where you want to go.

In the slewing mode different keys than those used in regular flight (Figure 15.4) are required. Pressing the T key several times in succession

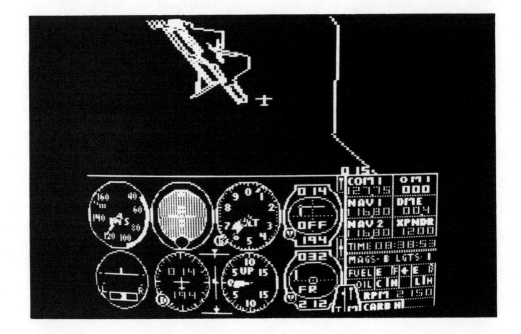

Figure 15.5 Arriving in Seattle, we overfly the airport.

causes the airplane to fly toward the north at jet speeds. Watch the figure in the left top display; it will rapidly increase. If the increase appears to be too slow, press F several times to speed it up. Once that number approaches 21300 press the G key to stop everything. Next press the F key several times; watch the number in the right display decrease. Again, if that's too slow, pressing T several times will increase the speed. When that number has decreased to 06500, press the G key again to stop the slewing action.

Return to the editing menu to get back into regular flight mode. Notice that there has been no change in the fuel quantity because no fuel consumption is registered in the slewing mode. Press the ESC key and change the 1 opposite SLEW to 0, press the ESC key again, and you'll be back in the regular flying mode. The OBI will have locked onto Seattle, and the DME will display the distance from the SEA VOR (Figure 15.5). Change the heading in the OBI until the CDI needle centers and take up the displayed heading to fly to the VOR.

Once there, you may want to fly around the area and look at the different airports and all the Puget Sound water before picking one of the airports and flying an approach.

CHAPTER SIXTEEN

Minimum En Route Altitudes and Other Matters

It occurred to me that it might be interesting to purposely crash our simulated aircraft into a mountain. To see what would happen, I took off from Van Nuys, California, flying outbound on the 030-degree radial from the VNY VOR, heading for Palmdale. Palmdale is not on the area chart provided with the simulator, but it is located on the intersection of the 053-degree radial from the Filmore VOR (FIM) and the 322-degree radial from the Pomona VOR (POM), roughly 33 nm northeast of Van Nuys (Figure 16.1).

Despite the fact that the minimum en route altitudes (MEAs) in that area are around 7,000 feet, I leveled out at 1,500 feet, and sat back, waiting to see what, if anything, might happen. Nothing did, probably because the program does not include any topographical features in that area. In fact, I eventually arrived at the location where the Palmdale airport would have been, still level at 1,500 feet. In real life this would be a good trick, because the elevation at Palmdale is 2,542 feet.

Next I turned right to a heading of 142 degrees and headed toward the POM VOR and Brackett Field. That route would take me over (or through) some rather respectable mountains, and the MEA for that route is a healthy 10,000 feet. After a while a black mess appeared in the distance ahead of the aircraft. I assumed it represented that mountain range. When I got close enough to see that it was just that, I was even able to see the highway that runs through Cajon Pass. I was still flying at 1,500 feet msl in the southern portion of the Majave desert, an area where the actual terrain elevation is in excess of 2,000 feet msl.

Eventually everything that was to be seen straight ahead turned solid black, and I assumed that any moment the program would respond with CRASH! It did not. I spent seven or eight minutes flying right through solid

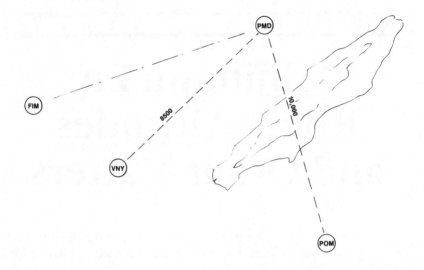

Figure 16.1 The route of flight from Van Nuys to Palmdale to Pomona while experimenting with flying through mountains.

rock. I finally emerged on the other side, with freeways and Bracket Field visible in the distance.

At this point I wondered if the same thing would happen if I attacked the mountain range from the other side, from *inside* the navigation area flying outward. I turned around and flew back toward the mountain, still level below 2,000 feet msl. Again I flew right through the rocks into no man's land.

The point is that the program does not necessarily pay attention to whether or not the airplane is at a flyable altitude. Therefore, it's up to us as pilots to remain aware of the terrain elevation in the areas over which we intend to fly. When we made that slewing flight from Chicago to Seattle, we'd have had to climb to 13,000 or 14,000 feet to clear the Rockies. There is no terrain elevation information on the area charts other than the altitude of the listed airports (the first of these two numbers listed below the name of the airport, the second number being the length of the longest runway in hundreds of feet—533-32 represents an airport elevation of 533 feet with a longest runway of 3,200 feet). We'll have to use the sectionals to determine safe cruising altitudes for our flights.

Don't ignore the importance of flying at safe as well as legal altitudes. If we consistently ignore the subject while flying the simulator, we're likely to develop bad habits that could be dangerous in the real world. The legal altitudes for VFR flights heading east, on courses between 001 and

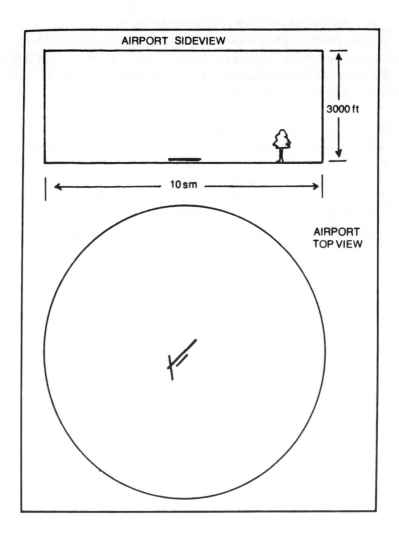

Figure 16.2 A diagram of an airport traffic area.

180 degrees, are odd thousands plus 500 feet (5,500, 7,500 etc.), and even thousands plus 500 heading west on courses between 181 and 360 degrees. In both instances, these figures are for cruise levels above 3,000 feet agl. For IFR flights the legal altitudes are odd thousands going east and even thousands going west, unless we're given instructions to the contrary by ATC.

There are no minimum-altitude restrictions when flying over unpopulated terrain. When flying over towns, cities or villages, a minimum altitude of 3,000 feet is in effect to prevent unnecessary noise pollution. There are also altitude restrictions when we're flying in the vicinity of airports with operating control towers. Such airports are surrounded by something known as *airport traffic areas*, circular areas ten statute miles in diameter, with the actual airport in the center (Figure 16.2). They extend from ground level to (but not including) 3,000 feet agl, and VFR aircraft are not permitted to fly through these airport traffic areas without being in contact with the control tower. Most airports with control towers are identified on the area charts by CT along with the primary tower frequency (CT 119.1), directly beneath the name of the airport.

There are many other restrictions about the airspace, such as terminal control areas (TCAs), the continental control area and so on, but there is no practical way to adhere to the regulations while flying the simulator. Once we are in an actual aircraft these regulations and restrictions become vitally important; novice pilots should make sure they are clearly understood before attempting to fly into or through areas other than uncontrolled airspace.

CHAPTER SEVENTEEN

The War Game

As an added attraction, the program disk includes a game described as "World War I Ace." I'm not much for video games in general and war games in particular, but since it's offered, we should take a closer look.

The game is a three-dimensional aerial battle in which you are expected to bomb enemy facilities while trying to shoot down opposing fighter aircraft and avoid being shot down yourself. The battlefield resembles a square checkerboard bordered by mountains on two sides and divided by what appears to be a river (Figure 17.1). On your side of the river are two airbases and a fuel depot, and on the opposing side are three factories, two airbases and two fuel depots.

You attempt to bomb the factories and fuel depots. This wouldn't be too difficult if it weren't for six enemy planes controlled by the program. You have no control over them, other than trying to position your own aircraft in such a way that you can destroy them by firing straight ahead.

In addition to the keys that control your aircraft in normal flight, four keys perform different functions in this mode. The W key is used to declare war. Pressing the X key drops bombs. Your aircraft is equipped with five bombs when the game begins. The space bar is your machine gun; it must be pressed repeatedly in quick succession in order to fire a burst of bullets. And, finally, the R key causes the scenery display to be replaced with a page of data showing the number of aircraft shot down, the number of successful bombing hits and so on. While this display is activated, all war action is held in abeyance. Pressing R again returns you to the battle field and the action continues.

To get into the war game mode, call up the editing menu (press the ESC key) and change the number opposite EUROPE 1917 from 0 to 1. Press the ESC key again and you'll be on the ground at the main airbase on your side of the river. Notice that the radio stack on the right side of the

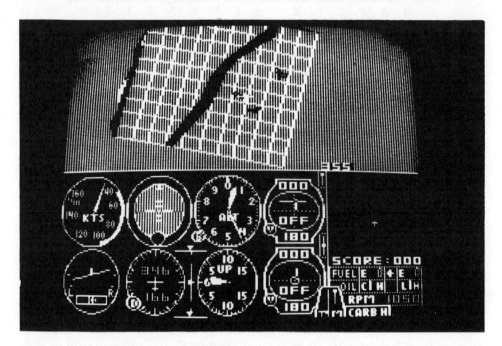

Figure 17.1 The battle field' used by the war game.

instrument panel has been replaced by a radar display, showing your own aircraft in the center. The area covered by this radar screen is approximately one square mile. You'll use this display constantly in your attempts to deal with the enemy fighters (Figure 17.2). You'll use the downward view (5 G) which includes a bomb sight to position yourself directly over the enemy targets you'll try to bomb.

Start by taking off in the usual manner. Once you are level at a few thousand feet, fly around the area to familiarize yourself with the location of the various targets. Use the forward view as well as the radar view to get the clearest possible picture. Since war has not as yet been declared (you must press W), you don't have to worry about interference from enemy aircraft. Once you're satisfied that you remember the lay of the land, return to your side of the river and declare war. Head toward the nearest target for your first bomb run, keeping an eagle eye out for some tiny light dots that will appear on your radar screen. They represent enemy fighters that must be avoided or shot down.

Assuming that you arrive safely over one of the enemy factories or

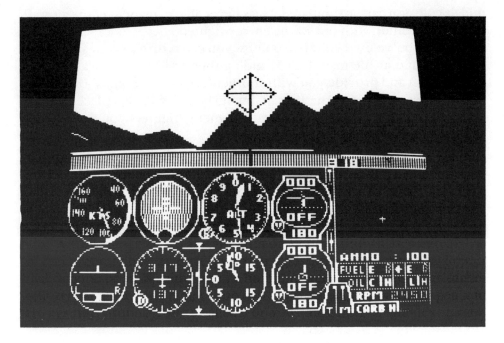

Figure 17.2 The forward view through the windshield in the war game mode.

fuel depots, press X to drop your bomb. A legend over the radar screen will tell you whether your drop was successful. If it wasn't, you can always turn around and try again, or you can go on to the next target, leaving the one you missed for another run. Don't try to bomb one of the enemy airbases. They are not considered targets by the program and you won't get credit even if it's a direct hit.

A word of warning. The aircraft you're flying either uses a great deal of fuel or it carries very little in its two tanks, as was true with the aircraft that were used in World War I. Watch the fuel quantity display; it takes only a short time for a tank to run dry. If it does, you'll suddenly begin to fall because your engine will quit. If you're quick about it, and if you were at a respectable altitude when the engine quit, you may be able to restart it by switching to the other tank (avoiding what is euphemistically referred to as ''uncontrolled contact with the ground'').

You'll find that shooting down enemy aircraft is quite a trick; your machine gun is most effective only when shooting straight ahead. It has a pretty good straight range, but poor side range. I don't know whether a midair collision can be simulated. I was never able to maneuver the aircraft to actually collide with one of the enemy fighters.

After you've used your five bombs, you can return to your side of the river and land at the main base. Landing there will automatically refuel your aircraft and provide you with another load of five bombs. Landing at the other base will refuel the aircraft, but no bombs will be replaced. While you're busy refueling at either base, any enemy fighters you managed to shoot down will be replaced.

If you've been hit by an enemy fighter, your aircraft will start to act strangely. You may be losing fuel or oil rapidly; in that event, try to get back to your side and land at one of the bases to refuel and effect whatever repairs are necessary.

The handbook that comes with *Flight Simulator II* includes a point system, assigning one point to each downed enemy aircraft, and 20 and 10 to each destroyed factory or fuel depot, respectively.

OTHER AVIATION SOFTWARE

Commercial Aviation Software
An Overview

Commercial aviation software for microcomputers is a relatively recent development. Only during the last two or three years have software manufacturers realized that there is a considerable market for such programs. Since then, we have seen rapidly increasing attendance by software manufacturers at major aviation conventions to exhibit their products to the pilot population.

Software designed for use in aviation falls into two distinct categories. One category involves programs designed to be used by fleet operators, corporate flight departments, commuter airlines, and air taxi and charter operators. They deal with all manner of sophisticated aviation-related data, such as passenger scheduling, aircraft maintenance schedules, pilot refresher training, accounting, and so on. Some are designed to run on very powerful computers and, usually carry hefty price tags. They are beyond the reach as well as beyond the need of the average private or business pilot. For that reason, I'll look at programs of this type only sketchily in this book.

The other category consists of programs designed for the average student, private or business pilot. They can be run on the standard microcomputers, Apples, IBM PCs or Commodores. These programs fall into two classes—those designed for instrument training and practice, and those intended to simplify the task of flight planning. Most of these programs are fairly inexpensive and easily within reach of anyone who can afford to own or rent an airplane.

In this part of the book we'll discuss some of the more popular aviation programs to give you an idea of their function and purpose. A list of the manufacturers of the programs discussed here, along with addresses and current prices, is included in one of the appendices.

Navlog
A Flight-Planning Program

Navlog is a computerized flight-planning program that incorporates unique and useful features. It is written for a 56K Apple II Plus or IIe equipped with the Microsoft CP/M SoftCard with two disk drives. The program package consists of two 5 1/4-inch disks: the Navlog program disk which must run in drive A, and the Navlog data disk, which must be in drive B. Also included is a booklet containing documentation. Before the program can be run, the CP/M system must be copied onto the program disk (COPY B: = A:/S).It is advisable to make at least one copy of both disks and to store the originals in a safe place to avoid some inadvertant mishap.

Once the CP/M system has been copied onto the program disk it will boot and immediately display the main menu that offers the user 11 choices:

NAVLOG MENU

1 = Open flight log   Current flight is:

2 = Enter new leg   Airplane is:

3 = Display/change leg   Your selection is:

4 = Insert leg
5 = Delete leg
6 = Display/print flight log
7 = Display/print airport file
8 = Change airplane
9 = Airplane data maintenance
10 = Airport data maintenance
11 = Weight and Balance

```
****************************************
*                                      *
*               NAVLOG                 *
*                                      *
*  Computerized Flight Planning System *
*                                      *
*     Copyright   1983    Flightware   *
*                                      *
****************************************
```

I - Flight plan

```
-----------------------------------------------------------------------------
!                           FLIGHT PLAN                                     !
!---------------------------------------------------------------------------!
!1.Type !2.Aircraft !3.Aircraft !4.TAS  !5.Departure !6.Departure !7.Cruise  !
!  IFR ! Ident     ! type      !  150  ! point      ! time       ! altitude !
!  VFR ! 5218T     ! CESSNA    !       ! SAF        ! prop  act  !          !
!  DVFR !           ! T 182 RG/R! Knots !            !            !  10500   !
!---------------------------------------------------------------------------!
!8.Route of flight                                                          !
!SAF - PHX (landing)                                                        !
!PHX - VNY (landing)                                                        !
!---------------------------------------------------------------------------!
!9.Destination           !10.Est. time enroute !11.Remarks                  !
!                        !   Hours    Minutes  !                            !
!  ABQ-PHX               !   See flight log     !                           !
!---------------------------------------------------------------------------!
!12.Fuel on board !13.Alternate airport !14.Pilots name.address !15.No aboard !
!  Hours Minutes  !                     !  Phone.A/C home base  !            !
!    6     46     !                     !                       !            !
!---------------------------------------------------------------------------!
!16.Color of aircraft !                                                     !
!                     !     Close flight plan with _____FSS        !
!   WHT/BRN           !                                                     !
-----------------------------------------------------------------------------
```

Figure 19.1 The FAA-type flight plan produced by the NAVLOG Computerized Flight Planning System.

As is so often the case, there is no "exit program" choice. To quit, one can simply remove the disks from the drives, but I always find that awkward.

To understand this menu, we must know the features that make up the program. Stored in the program are the data associated with some 100 airports and VORs. There is room for the user to enter up to 900 additional airport or VOR data. Also stored in the program are the data for up to 64 aircraft, used by the program to calculate time en route, fuel required, weight and balance and other data related to a given flight leg.

The booklet that is part of the program package gives a fairly clear description of the options available to the user. Sadly, it (as well as the program itself) contains numerous typographical errors and some strange syntactical forms, such as using lowercase letters for VORs.

```
II  -  Flight log

FROM: SAF
TO                    ALT    DIST   M-C  WCA       NOTES           G-S   FUEL   ETE
----------------     -----  -----   ---  ---  ------------------   ---   ----   -----
PHX                  10500  314.5   232   -1  ------------------   132   33.0   142.6
Landing                     ------                                       -----  -----
                            314.5                                        33.0   142.6

VNY                   8000  332.1   264   -4  ------------------   134   34.4   149.1
Landing                     ------                                       -----  -----
                            332.1                                        34.4   149.1

Totals                      ======                                       =====  =====
                            646.6                                        67.4   291.7
                                                                              ( 4:52)
```

Figure 19.2 The second page prints the details involved with a flight from Santa Fe, New Mexico to Van Nuys, California with a stop in Phoenix, Arizona.

The first choice clears the screen and replaces the main menu with another list of choices, such as listing or printing all flights on file, entering a new flight, selecting a previously entered flight, deleting a flight, or changing the name of a flight. Since flight legs are stored by number, the computer must be given that number. It is useful to list all flights on file so you will select the right number.

The second choice from the main menu is used to enter new leg data. The new leg will automatically become part of the current numbered flight. When the identifiers for the departure and destination points have been entered the program automatically computes the true course and distance if the airports or VORs representing departure and desintation points are stored on the disk. The syntax for entering identifiers is three letters for VORs and three letters followed by a period for airports (SAF and SAF.). After the user has keyed in altitude and wind data, the program computes and displays true heading, magnetic heading, ground speed and time en route in minutes, and fuel requirements.

The third, fourth and fifth choices are self-explanatory. The sixth choice prints a flight plan and all data associated with the selected flight, as shown in Figures 19.1 through 19.3. The seventh choice displays or prints all stored airport and VOR data. Choice number eight is used to select the type of aircraft you'll be flying. Numbers nine and ten are used to change previously recorded airport, VOR or airplane data. The last choice must be used before a flight to enter the pilot, passenger, baggage and fuel weight data applicable to a given flight.

```
III  -   Frequency reference

                          TWR/VOR         GROUND          ELEV.
                          -------         ------          -----
SANTA FE                  110.600                         6260

PHOENIX                   115.600                         1180

VAN NUYS                  113.100                          810




IV   -   Weight and Balance

    Pilot      =    135 Lbs               Gross wt = 2833 Lbs

 F Passenger =    160 Lbs     Total moment/1000 = 112   lb-in

 R Passenger =      0 Lbs                     C/G =   40   in

 R Passenger =      0 Lbs

    Fuel      =    528 Lbs

  Baggage 1  =     60 Lbs

  Baggage 2  =      0 Lbs



 V   -   Flight Cost = $ 121.54
```

Figure 19.3 The third page lists frequencies, airport elevations (the one for Santa Fe is wrong, it's 6344), weight and balance data, etc.

The program does not take airways, restricted areas and such into consideration. It simply determines flight-related data based on a straight line between the two points entered by the user. Flights must be entered in the form of individual legs, each a direct route between two points. Flights may consists of as many as 22 legs. While this may seem tedious, it only has to be done one time, as the computer will store these data on the data disk, from which they can be recalled the next time the flight is to be flown.

Using
Jerry Kennedy's
Flight Planner

The flight planning program produced by Jerry Kennedy is designed to run on an Apple II, II Plus or IIe with a minimum of 48K and one disk drive. It has a number of unique features and, except for the fact that it's very slow (it is written in Applesoft Basic), it appears to be useful.

Briefly, here is what it does. You key in your departure and destination and the altitude above ground level at which you wish to cruise, and the program produces all the nav aid data for two routes. It then asks for wind and true airspeed data and produces ground speed and time en route for the two routes you've selected. The advantage of having the program produce two routes is that one or the other might be impractical because of thunderstorm activity or other weather problems.

If your version of the program includes the area navigation option, it asks whether you want the RNAV route computed. In each case, you've been given the option of having the program compute the return flight data and have the results sent to the line printer.

When the disk is booted for the first time, the program asks questions about your printer. These configure the program to conform to the printer you're using. During subsequent runs, these questions are skipped.

When the program is used for flight planning purposes, it starts by asking:

```
ENTER APPROXIMATE ALTITUDE AGL
```

If you key in an altitude that is too low to permit signal reception from the en route VORs, the program automatically increases the altitude in 1,000-foot increments until it finds a useable reception altitude. (To find the lowest possible reception altitude, key in 0). The next input request is:

```
POINT OF DEPARTURE
STATE
```

Use the three-letter identifier for the point of departure and the two-letter postal state identifier for the state. The program will automatically change it to the full name of the state. This is followed by:

```
DESTINATION
STATE
CORRECT ENTRY (Y/N)?
```

which should be treated in the same way. Next we see:

```
FIRST VOR
STATE
LAST VOR
STATE
CORRECT ENTRY (Y/N)?
```

which is self-explanatory. In a moment, the following line appears at the top of the screen:

```
STATES WHICH BORDER ROUTE OF FLIGHT:
```

Then comes a rather long wait during which the display tells us the program is searching the nav aid data files. Eventually the list of states appears. I don't really know what practical purpose this list serves; perhaps it's just to have something happening while the computer is producing the two routes of flight. Depending on the distance to be covered, the waiting time varies. The program can compute very long flights, such as from LAX to JFK, but it will take several minutes. The computer then displays the two routes and the distance for each in nautical miles.

```
WOULD YOU LIKE TO COMPUTE FLIGHT PLAN? (Y/N)
```

is the next question. If you answer in the affirmative, you're asked to key in:

```
WIND DIRECTION
WIND VELOCITY
TRUE AIRSPEED
```

```
WIND DIRECTION 240

WIND VELOCITY   12

TRUE AIRSPEED   145

ALTITUDE (AGL) 6000


ID  FREQ  TO ID  FREQ  FR  TO  NM  TRIP


SAF        TO SAF 110.6 153 153    5  0:02

SAF 110.6 TO ABQ 113.2 218 218   47  0:23

ABQ 113.2 TO SJN 112.3 240 237  121  1:17

SJN 112.3 TO PHX 115.6 233 232  149  2:24

PHX 115.6 TO PHX       259 259    6  2:27


TOTAL MILES = 328 NM

GROUNDSPEED = 133 KT


STATES WHICH BORDER ROUTE OF FLIGHT

TEXAS

NEW MEXICO

ARIZONA
```

Figure 20.1 The printout of a flight log from Santa Fe to Phoenix, using the airways (VOR to VOR).

The wind velocity and TAS must be entered in knots.

Once those data have been keyed in, the program produces a flight plan like the one shown in Figure 20.1. Finally, you're offered the option of having all this information printed by your line printer, after which you may ask for the return route and/or the RNAV routing, one of which is shown in Figure 20.2.

```
FIX (VOR FREQ    BRG / DIST) TO   FR   NM TRIP

SAF (                    )        153      0:00

SAF (SAF 110.6   0.0/  0.0)  153 235    5 0:02

# 2 (ABQ 113.2 360.0/ 16.8)  235 235   35 0:18

# 3 (SJN 112.3 360.0/  3.5)  232 232  128 1:15

PHX (PHX 115.6   0.0/  0.0)  231 259  151 2:23

PHX (                    )   259       6 2:26


   DISTANCE = 325
```

Figure 20.2 This printout shows the same route, this time computed for area navigation showing the waypoint bearing and distance from the referencing VOR/DME.

Taking the print-outs of this program along on a flight is a great convenience; it eliminates a lot of fighting with charts and searching for frequencies in a cramped cockpit. The program contains the data for all VORs in the contiguous 48 states, and users can update the stored data when frequency, identifier or other changes make it obsolete. The easiest way to update data is to send the disk to the manufacturer along with $10. The other is a do-it-yourself method described in detail in the instruction booklet. Caution: before attempting to update the nav aid or airport files, be sure to make a copy of the disk, using the COPYA program from the Apple System Master disk. There's always the danger of accidentally erasing an entire data file; I hate to think of the time necessary to type in all these data from scratch.

CHAPTER TWENTY-ONE

A Family of Aviation Software from Flight Opps

This group of programs for the IBM PC is designed primarily for use by operators of high-performance corporate aircraft, airtaxi and charter operators, and commuter airlines. Included in the group are the *The Flight Planner*, *Flight Utilities*, *The Weather Station* and *PT6 Engine Monitoring*. The following is a brief description of each.

Flight Planner

This package is a group of subprograms designed to help plan fuel loads, compute direct routes of flight and determine the most advantageous cruising altitude. It produces a flight plan report that summarizes the navigation, performance and fuel requirements for any flight. This information is computed from user entries outlining the flight, combined with information from aircraft, airport, nav aid and weather data bases. The program is compatible with all commercial weather data base services that produce FAA format winds-aloft reports receivable as text files. It can be used with all aircraft types and permits user-customized data bases for each aircraft.

When used in the area navigation mode, the program produces the RNAV routing and determines the waypoints, all evenly spaced. It prints all waypoint data such as latitude, longitude, and VOR offset information ready for input into the on-board RNAV equipment.

Flight Utilities

This is a collection of ten programs that perform many frequently-needed preflight calculations. When given two station identifiers it computes the great circle distance or the great circle course outbound from the first identifier. Add ground speed and it computes time en route. It uses

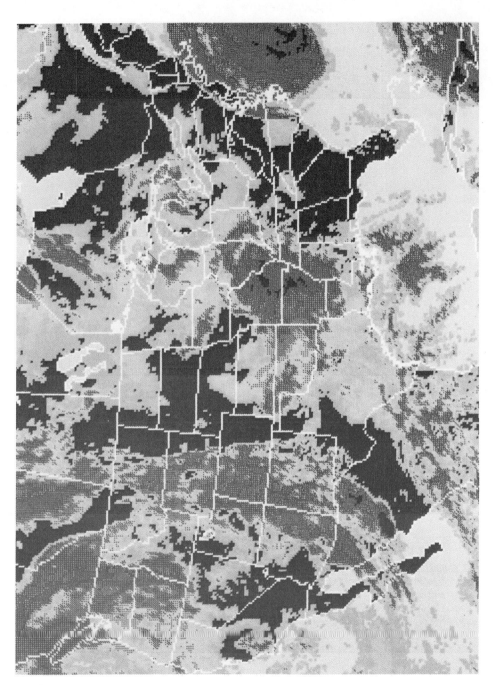

Figure 21.1 A computer generated printout of the weather systems covering the entire U.S. except for portions of Alaska and Hawaii.

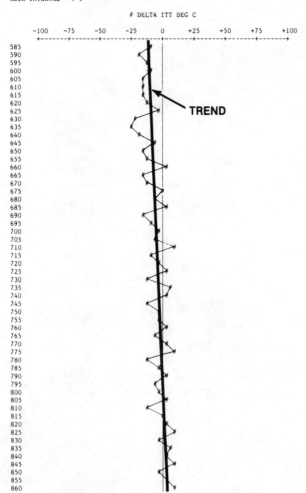

A sample graph showing trends of turbine temperature. Each data point represents the difference of the actual temperature from the normal. The graph has been reduced to 50% of the original and lines have been drawn in to connect points for clarity. The bold straight line has been drawn in to show the trend of the readings. Small deviations between readings are normal. Long term trends are most important, especially when compared with the associated graphs for turbine speed and fuel flow. Graphs should be interpreted by maintenance personnel familiar with engine trend monitoring. For more information, consult the Pratt & Whitney Engine Trend Monitoring service letter for your engine model.

Figure 21.2   A printout produced by the engine trend monitoring system.

calibrated airspeed, altitude and temperature to produce true airspeed, or it can use true airspeed, altitude and temperature to return the mach number. Given pressure altitude and temperature, it computes the density altitude. When aircraft course, airspeed, wind direction and velocity are keyed in, it computes the wind correction angle and ground speed. Using a station identifier and the difference in time from Greenwich Mean Time, it displays morning twilight and sunrise times as well as evening twilight and sunset times. Given true airspeed, course and the wind correction angle, it computes wind direction and velocity. And finally, it can compute the crosswind component using course, wind direction and velocity.

The Weather Station

This program converts the IBM PC into a remote weather terminal. It performs the communications and graphics processing to download weather reports and color graphics, and it will reproduce weather graphics on a dot-matrix printer (see Figure 21.1). The program communicates with the Weather Data Bank of WSI Corporation in Bedford, Massachusetts over telephone lines. It can be operated in two modes, manual or automatic. In manual mode, the user interacts with the WSI computer through the keyboard, issuing commands and viewing the information as it is received. In automatic mode the user may set up weather requests before running the program and leave during the time it takes to receive the information. The program will store all incoming data in a standard text file for display or printing at a later time. Furthermore, it can be programmed to call and receive weather information hours later or at regular intervals.

PT6 Engine Trend Monitoring

This program performs the correction calculations required by the Pratt & Whitney procedures for engine trend monitoring and provides a means to record engine readings, and produce reports and graphics of engine trend data. When analyzed by qualified maintenance personnel, engine trend graphs can reveal potential problems at very early stages and corrective action can be taken before the problem becomes more serious (Figure 21.2).

CHAPTER TWENTY-TWO

The RNAV3 Navigator From Briley

Briley Software of Livermore, California, produces and markets a program called *RNAV3 Navigator*. This program computes flight paths via direct routes (area navigation) or via the Victor Airways. It is designed to run on a Commodore Pet/CBM or Commodore 64 microcomputer with 16K. It is available in two versions; one covers the 11 Western states, and the other covers the 17 Northeastern states. A version for the Central and Southern states is in the planning stages.

The program computes straight-line courses between any two points within the areas covered by its data base for aircraft equipped with DME. Each waypoint, the three-letter identifier, the frequency of each applicable VORTAC or VOR/DME as well as radial and distance can be displayed on the screen or sent to the line printer. It also displays or prints the magnetic course, nautical miles traveled and nautical miles left to destination at each waypoint.

For aircraft without DME, it performs the same functions using VOR-to-VOR airway routing, displaying and/or printing all the appropriate nav aid data.

When activated, the program asks the pilot to key in the departure and destination coordinates and, in the case of the area navigation option, the desired distance between waypoints. Once these data have been entered, the program performs all the necessary trigonometric calculations to produce the required results.

For a typical flight, using the area navigation option, the display would look something like this:

```
RNAV3 NAVIGATOR (WEST)      OPTIONS: D
DEPARTURE      LAT:   37.42 LON:    121.5
DESTINATION LAT:    36.4   LON:    121.37
TRUE COURSE = 170 DEG   62.9 NM


WP------VORTAC-----DIST          MC   NM   REM
   NAME/FREQ/RADIAL  (NM)
1  SJC  114.1   14.9 12.1   153  10   52.9
2  SJC  114.1  116.3 13.7   153  20   32.9
3  SNS  117.3  333.5 13.2   153  20   12.9
4  SNS  117.3  268.5    .7   153  13    0
```

CHAPTER TWENTY-THREE

The A-3
Airplane Simulator

Mind Systems Corporation of Northampton, Massachusetts, has developed the *AirSim-3 Airplane Simulator* for the family of Apple II and III microcomputers with 48K of RAM. It simulates a generic light single-engine aircraft, and includes the basic flight instrumentation found in light aircraft plus the navigation instrumentation required for instrument flying practice. There is also a mode that lets the user enter wind conditions, runway locations and the locations of navigational aids at his or her discretion. Included is a move-with-joystick mode for those users whose computers are so equipped.

The program includes some rather sketchy scenery representing the California coast line (see Figure 23.1), from north of San Francisco to south of San Diego, and a fictitious "home" scene located somewhere north of Bakersfield. This latter is believed to be an aid for beginning pilots. It includes three runways for practice and recreational flying.

In addition to the "home" airport, the program includes realistic representations of the airports at San Francisco, Monterey, Santa Barbara, Los Angeles, Long Beach and San Diego. Each of these corresponds to the actual runway patterns and is equipped for instrument approaches (see Figure 23.2). Also included are all VORs and NDBs located in the area covered by the program, making it possible to take off from any airport and fly on instruments to any other airport. To avoid the tedium that can result from trying to fly long distances at realistic speeds, the user can use an option that moves the airplane quickly from any point to any other point within the limits of the navigational area.

The program includes a radar view that lets the user see the airplane relative to the chart-like representation of the ground below. It also shows the locations of the nav aids in the area to which the navigation radios are tuned.

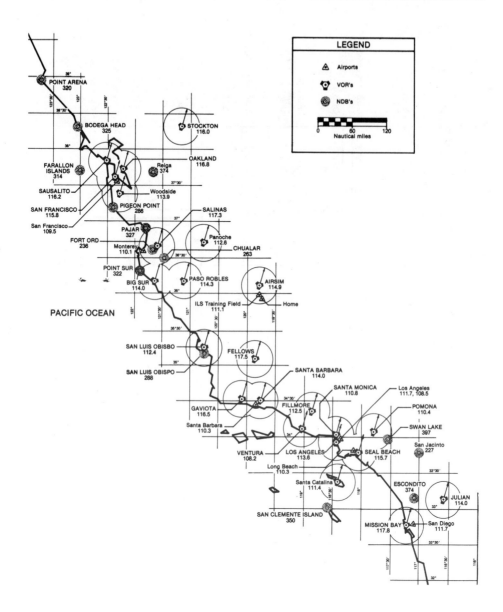

Figure 23.1 A chart showing the area covered by the program and the airports and navaids that are included.

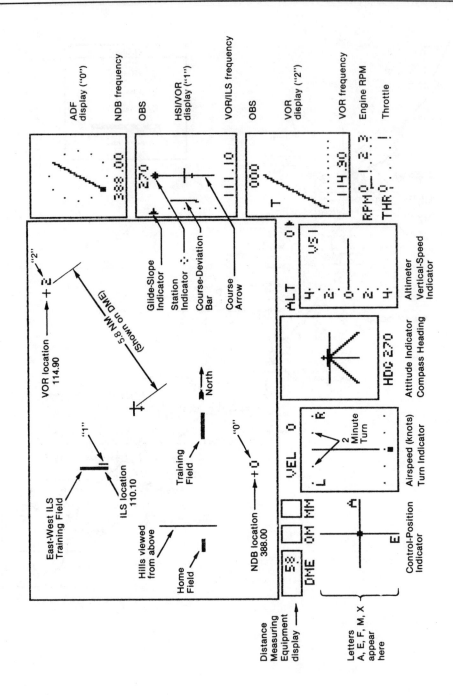

Figure 23.2 An explanation of what is available in the display.

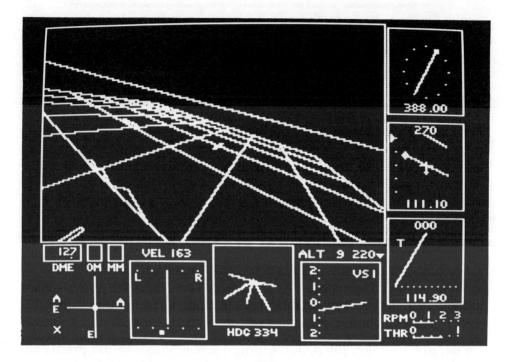

Figure 23.3 The display shows some scenery and the basic instruments, the representation of which lacks realism in design.

On computers equipped with Mockingboard Sound II or Sound/ Speech I boards, the program produces engine sounds and the wheel-screech upon landing.

The program is well-thought-out and appears to perform as promised, but I object to the instrument representations. They don't resemble the real thing (Figure 23.3).

Also available from Mind Systems is a Spitfire simulator that simulates the Battle of Britain in 1941. It includes a number of enemy aircraft of different types: Messerschmitt 109 fighter, Focke-Wulf 190 fighter, Stuka dive bomber, Heinkel 111 bomber, Junker 52 tri-motor transport, a Zeppelin dirigible and two unidentified flying objects. The simulated Spitfire is the Supermarine Spitfire Mk1 equipped with a 1030 horsepower Rolls Royce Merlin II engine, capable of level flight at up to 360 mph. The program is available for the Apple II, II +, and IIe, with at least 48K RAM, 1 disk drive (DOS 3.2 or 3.3), game paddles or joystick and Applesoft or equivalent in ROM.

CHAPTER TWENTY-FOUR

The Air Nav Workshop

Space-Time Associates of Manchester, New Hampshire, is represented in the aviation microcomputer simulator field by a program called *Air Nav Workshop* that is intended as an instrument training aid for student pilots. It operates on the Apple II, II+, IIe and III with at least one disk drive (DOS 3.3), 48K and Applesoft in ROM. A second disk drive simplifies the operation of the program by eliminating the need for disk swapping. A line printer with a Grappler+ or comparable interface card can be used to dump the high-resolution screen image to the printer to examine previous flights.

The program is described by its author, Ken Winograd as a ''teaching/learning tool for aviation enthusiasts.'' It is designed to acquaint users with the radio navigation environment. The program may be customized by creating navigation areas that could represent the home environment or any environment on earth. Customized areas may be saved on a second data disk to be used over and over again. The navigation aids available with the program are VORs and NDBs. The simulated aircraft is equipped with dual VORs and a single ADF as well as an airspeed indicator.

The *Air Nav Workshop* consists of four separate programs: ADF/NDB Training Aid, VOR/VOR Training Aid, Navigation Simulator and Area Personalization. The first two acquaint novice pilots with the way the onboard navigation radios interface with the ground-based VORs and NDBs. The third program represents an open-ended simulation. The user controls airspeed, heading and all navigation radios. Wind direction and velocity may be keyed in at the user's discretion. The fourth program is used to customize the navigation area. This program makes it possible to design as many different areas as desired or several versions of the same area that use diferent nav aids. All of these can be saved on the same data disk.

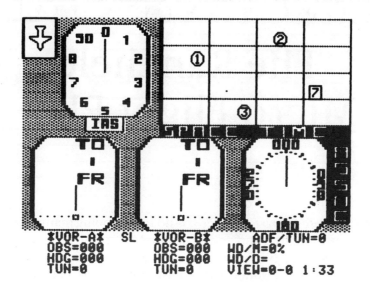

Figure 24.1 The display lay-out: the navigation area covers the upper right quarter.

The documentation is well-written, but is reproduced directly from a single-spaced, right-justified, dot-matrix print-out which makes it difficult to read.

When the program is activated, the first two displays (the title page and the main menu), are annoying in their pretentiousness. The graphics used are impressive, but all the elaborate design work gives the impression that the program is a game rather than a serious instructional tool. Everything seems to work as it's supposed to, but the instrument representations lack realism, and it takes a bit of practice to figure out how to control the nav radios, heading, etc. There is another extraneous graphic feature consisting of a little airplane symbol that cavorts across the screen for no apparent reason.

The upper-right quarter of the screen (Figure 24.1) represents the navigation area in which the aircraft's flight path can be observed. Certain key strokes may be used to take a closer look at sections of this area in an enlarged format.

Unlike some other flight simulation programs, this one does not teach you how to control the aircraft. It is strictly an navigation instrument trainer, and as such, is really quite good.

CHAPTER TWENTY-FIVE

The Ranchele Micro Flight Plan

The *Ranchele Micro Flight Plan* is no longer being manufactured, but since it is exceptionally good and large numbers of the program are in use (and may still be available from some sources), I include it here.

The program is unique in that customization is performed by the manufacturer. The purchaser receives a lengthy questionnaire that must be completed. It includes the weight and balance specifications and performance data applicable to the aircraft with which the program is to be used. The data are keyed into the program by Ranchele. Since Ranchele is no longer marketing the program, this may present a problem for new users, but the program is written in BASIC, and can be listed on a line printer. A programmer should then be able to study the listing and key in the applicable data.

The program runs on the Apple II family of computers, requiring a minimum of 48K, two disk drives and a line printer. Drive A contains the actual program disk, and drive B uses a disk that contains routes entered by the user. The program disk includes three demonstration routes that can be used initially so new users can learn how the program functions.

When activated, the program offers the user four choices:

```
1.  INPUT MASTER ROUTE SEGMENTS
2.  PREPARE FLIGHT LOG ONLY
3.  PREPARE FLIGHT LOG AND PLAN
4.  EXIT PROGRAM
```

The first selection is used to key in flight route data for frequently flown trips; the data are saved on disk for future use. The second choice computes and prints a flight log, and should be used when there is no need for the computer to produce an actual FAA-type flight plan. The third is

```
                              FLIGHT LOG
                               N310 PG
                              -----------

                 WASHINGTON DULLES-INDIANAPOLIS
                                                TIME OFF:
                                                          -------
                                   M I L E S    T I M E    S P E E D
                                   ---------    -------    ---------
       ROUTE DATA             HDG  LEG  TO GO   LEG TO GO  EST    ACT
=================================================================================
WASHINGTON DULLES-MARTINSBURG  332   32   401   0:17  2:48  110
DIR   MGW-111.6    MEA 5000
---------------------------------------------------------------------------------
MARTINSBURG-MORGANTOWN         284   94   307   0:39  2:09  143
V44   MGW-111.6    MEA 5000
---------------------------------------------------------------------------------
MORGANTOWN-ZANESVILLE          289   97   210   0:40  1:29  144
V144  ZZV-111.4    MEA 4000
---------------------------------------------------------------------------------
ZANESVILLE-APPLETON            295   34   176   0:14  1:15  146
V144  APE-116.7    MEA 3000
---------------------------------------------------------------------------------
APPLETON-DAYTON                267   84    92   0:36  0:39  140
V12   DAY-114.5    MEA 3000
---------------------------------------------------------------------------------
DAYTON-INDIANAPOLIS            264   92     0   0:39  0:00  140
V50   IND-116.3    MEA 3000
=================================================================================
TOT DISTANCE: 433   TOT TIME: 3:05    AVER SPEED: 140   WD ALTITUDE: 6000
% POWER: 65         AIRSPEED: 158     FUEL FLOW: 97#    TOT FUEL: 321#/$107
=================================================================================
WIND STATION        3000       6000        9000         ALTITUDE/TIME/FUEL$$

                    230/15     240/20      240/30         3000   3:04   $105
                    230/15     240/20      240/30         6000   3:05   $107
                    230/15     240/20      240/30         9000   3:12   $113

FUEL COST/GALLON: $2.00
```

Figure 25.1 The flight log for the first leg of a flight from Washington via Indianapolis and Kansas City to Denver.

used if a flight plan is to be printed in addition to the flight log. The last choice is self-explantory.

Since it is written in BASIC, the program is slow, but performs an unblievable number of computations. Among other functions, it produces a print-out that includes more detailed information about any proposed flight than the average pilot would ever put together himself. Possibly more data than are actually needed are provided.

```
                        WASHINGTON DULLES-INDIANAPOLIS


==================================================================================
X      X        X CESSNA      X         X DEPARTING-      X ZULU-   X            X
X VFR  X N210 PGX             X 158KTS  X                 X         X 6500 FT    X
X      X        X T210/A      X         X WASHINGTON DULLES                      X
X 1330                                                            X           X
----------------------------------------------------------------------------------
X ROUTE OF FLIGHT-                                                              X
X  DIR MARTINSBURG  V44 MORGANTOWN  V144 APPLETON                               X
X  V12 DAYTON  V50 INDIANAPOLIS                                                 X
X                                                                               X
----------------------------------------------------------------------------------
X DESTINATION-  X TIME-  X   REMARKS-                           X FUEL-   X
X               X        X                                      X         X
X INDIANAPOLIS  X 3+05   X                                      X 4+43    X
----------------------------------------------------------------------------------
X ALTERNATE-          X   PAUL GARRISON          X       X                      X
X                     X   SAF                    X 2 SOULS X GREEN & BEIGE      X
X                     X   505 982 0873           X       X                      X
==================================================================================
X  FSS 776-2727     FORCAST 776-1640     WX 429-0337        WX 639-4200     X
X                                                                           X
X GR TO WT: 3285#  LDG WT: 2964#       FUEL ON LDG: 159#   TIME REMAIN: 1+38 X
==================================================================================

GR TO WT INCLUDES: 2 PASS- 270#  BAGGAGE- 25#  FUEL- 480#  EMPTY WT- 2510#.
ALITITUDE INCREASED TO REFLECT MEA - PLEASE DOUBLE CHECK
TIME TO REACH ALTITUDE: 11 MINUTES - AVERAGE RATE OF CLIMB: 590 FEET/MINUTE




WEIGHT & BALANCE APPEARS OK BUT PLEASE CHECK
       TAKE-OFF: 3285# - 138.1 LBS-INS/1000
       LANDING: 2964# - 124.3 LBS-INS/1000


                                  ZULU          LOCAL
                                  ----          -----
         DEPART WASHINGTON DULLES  1330      8:30 AM EST
         ARRIVE INDIANAPOLIS       1635     10:35 AM CST
```

Figure 25.2 The FAA-type flight plan for the first leg of the flight.

The program does not display the flight log and/or flight plan data on the screen, but automatically produces a print-out (see Figures 25.1 through 25.6). The line printer must be ready before the program can be used. The printed flight log is particularly well designed; it includes space for entering actual in-flight data that can then be compared to those produced by the program.

```
                              FLIGHT LOG
                               N210 PG
                              ----------
                       INDIANAPOLIS-KANSAS CITY
                                                    TIME OFF:
                                                    --------
                                  M I L E S      T I M E     S P E E D
                                  ---------      -------     ---------
       ROUTE DATA            HDG   LEG  TO GO   LEG  TO GO  EST    ACT
======================================================================
INDIANAPOLIS-TERRE HAUTE     244   45    340    0:24  2:28  112
V50    HUF-111.8    MEA 3000
----------------------------------------------------------------------
TERRE HAUTE-DECATUR          280   76    264    0:34  1:54  134
V50    DEC-117.2    MEA 2500
----------------------------------------------------------------------
DECATUR-CAPITAL              282   37    227    0:17  1:37  134
V50    CAP-112.7    MEA 2400
----------------------------------------------------------------------
CAPITAL-QUINCY               265   76    151    0:33  1:04  137
V50    UIN-113.1    MEA 3000
----------------------------------------------------------------------
QUINCY-MACON                 254   57     94    0:24  0:40  140
V116   MCM-112.9    MEA 2700
----------------------------------------------------------------------
MACON-KANSAS CITY            252   94      0    0:40  0:00  140
V116   MKC-112.6    MEA 3000
======================================================================
TOT DISTANCE: 385   TOT TIME: 2:52     AVER SPEED: 134   WD ALTITUDE: 6000
% POWER: 65         AIRSPEED: 158      FUEL FLOW: 97#    TOT FUEL: 300#/$100
======================================================================
WIND STATION        6000        9000        12000        ALTITUDE/TIME/FUEL$$

                    300/25      310/30      330/40        6000   2:52  $100
                    300/25      310/30      330/40        9000   2:51  $100
                    300/25      310/30      330/40        12000  2:46  $101

FUEL COST/GALLON: $2.00
```

Figure 25.3 The flight log for the second leg of a flight from Washington via Indianapolis and Kansas City to Denver.

The documentation, though occasionally a bit redundant, is very clear and thorough. The displays produced by the program ask for each input in easy-to-understand phrases, showing sample input where the format is important. There are a sufficient number of error-trapping routines, such as ABOVE OK? (Y/N), to give the user a chance to correct input errors.

```
                           KANSAS CITY-DENVER

=================================================================================
X      X          X CESSNA      X           X DEPARTING-     X ZULU-   X             X
X VFR  X N210 PGX               X 166KTS     X               X         X 10500 FT  X
X      X          X T210/A      X           X KANSAS CITY    X 2042    X             X
---------------------------------------------------------------------------------
X ROUTE OF FLIGHT-                                                                X
X  V10N TOPEKA  V4 DENVER                                                         X
X                                                                                 X
X                                                                                 X
---------------------------------------------------------------------------------
X DESTINATION-  X TIME-  X   REMARKS-                              X FUEL-       X
X               X        X                                        X             X
X DENVER        X 2+46   X                                        X 4+35        X
---------------------------------------------------------------------------------
X ALTERNATE-            X  PAUL GARRISON          X        X                    X
X                       X  SAF                    X 2 SOULS X GREEN & BEIGE      X
X                       X  505 982 0873           X        X                    X
=================================================================================
X  FSS 776-2727    FORCAST 776-1640    WX 429-0337        WX 639-4200          X
X                                                                                 X
X  GR TO WT: 3285#   LDG WT: 2982#      FUEL ON LDG: 177#    TIME REMAIN: 1+49  X
=================================================================================
```

GR TO WT INCLUDES: 2 PASS- 270# BAGGAGE- 25# FUEL- 480# EMPTY WT- 2510#.
TIME TO REACH ALTITUDE: 17 MINUTES - AVERAGE RATE OF CLIMB: 617 FEET/MINUTE

WEIGHT & BALANCE APPEARS OK BUT PLEASE CHECK
        TAKE-OFF: 3285# - 138.1 LBS-INS/1000
        LANDING: 2982# - 125   LBS-INS/1000

MULTIPLE FLIGHT SUMMARY:

```
        WASHINGTON DULLES-INDIANAPOLIS 433NM   3:05    321#   $107
        INDIANAPOLIS-KANSAS CITY        385    2:52    300    100
        KANSAS CITY-DENVER              492    2:46    303    101
                                       --------------------------------
        TOTAL                          1310NM  8:43    924#   $308
                                     . ==============================
```

```
                                  ZULU        LOCAL
                                  ----        -----
        DEPART WASHINGTON DULLES   1330     8:30 AM EST
        ARRIVE INDIANAPOLIS        1635    10:35 AM CST
        DEPART INDIANAPOLIS        1705    11:05 AM CST
        ARRIVE KANSAS CITY         1957     1:57 PM CST
        DEPART KANSAS CITY         2042     2:42 PM CST
        ARRIVE DENVER              2328     4:28 PM MST
```

Figure 25.4 The FAA-type flight plan for the second leg of the flight.

```
                              FLIGHT LOG
                               N210 PG
                              -----------
                         KANSAS CITY-DENVER
                                                    TIME OFF:
                                                    -------
                              M I L E S     T I M E      S P E E D
                              ---------     -------      ---------
        ROUTE DATA            HDG   LEG  TO GO   LEG  TO GO  EST    ACT
===================================================================================
KANSAS CITY-TOPEKA            260    50   442   0:25  2:21  118
V10N  TOP-117.8   MEA 3000
-----------------------------------------------------------------------------------
TOPEKA-SALINA                 254    98   344   0:31  1:50  191
V4    SLN-115.3   MEA 5000
-----------------------------------------------------------------------------------
SALINA-HILL CITY              275   124   220   0:40  1:10  188
V4    HLC-113.7   MEA 5500
-----------------------------------------------------------------------------------
HILL CITY-GOODLAND            266    69   151   0:22  0:48  190
V4    GLD-115.1   MEA 5500
-----------------------------------------------------------------------------------
GOODLAND-THURMAN              273    73    78   0:23  0:25  188
V4    TXC-112.9   MEA 7000
-----------------------------------------------------------------------------------
THURMAN-DENVER                263    78     0   0:25  0:00  190
V4    DEN-117.0   MEA 8000
===================================================================================
TOT DISTANCE: 492  TOT TIME: 2:46     AVER SPEED: 177    WD ALTITUDE: 12000
% POWER: 65        AIRSPEED: 166      FUEL FLOW: 97#      TOT FUEL: 303#/$101
===================================================================================
WIND STATION        6000      9000         12000        ALTITUDE/TIME/FUEL$$

                    50/10     60/15       65/25           6000   3:03  $105
                    50/10     60/15       65/25           9000   2:56  $103
                    50/10     60/15       65/25          12000   2:46  $101

FUEL COST/GALLON: $2.00
```

Figure 25.5 The flight log for the third leg of a flight from Washington via Indianapolis and Kansas City to Denver.

I would say this is one of the better programs of its type. It seems only logical that some software distributor might revive it.

```
                       INDIANAPOLIS-KANSAS CITY

================================================================
X      X       X CESSNA      X          X DEPARTING-   X ZULU-   X         X
X VFR  X N210 PGX            X 158KTS   X              X         X 6500 FT  X
X      X       X T210/A      X          X INDIANAPOLIS X 1705    X          X
----------------------------------------------------------------
X ROUTE OF FLIGHT-                                                         X
X  V50 QUINCY                                                              X
X  V116 KANSAS CITY                                                        X
X                                                                          X
----------------------------------------------------------------
X DESTINATION-   X TIME-  X   REMARKS-                        X FUEL-   X
X                X        X                                   X         X
X KANSAS CITY    X 2+52   X                                   X 4+43    X
----------------------------------------------------------------
X ALTERNATE-        X  PAUL GARRISON       X        X                  X
X                   X  SAF                 X 2 SOULS X GREEN & BEIGE    X
X                   X  505 982 0873        X        X                  X
================================================================
X  FSS 776-2727    FORCAST 776-1640    WX 429-0337      WX 639-4200     X
X                                                                       X
X GR TO WT: 3285#  LDG WT: 2985#     FUEL ON LDG: 180#  TIME REMAIN: 1+51 X
================================================================
```

GR TO WT INCLUDES: 2 PASS- 270# BAGGAGE- 25# FUEL- 480# EMPTY WT- 2510#.
TIME TO REACH ALTITUDE: 11 MINUTES - AVERAGE RATE OF CLIMB: 590 FEET/MINUTE

WEIGHT & BALANCE APPEARS OK BUT PLEASE CHECK
        TAKE-OFF: 3285# - 138.1 LBS-INS/1000
        LANDING: 2985# - 125.2 LBS-INS/1000

MULTIPLE FLIGHT SUMMARY:

```
        WASHINGTON DULLES-INDIANAPOLIS 433NM   3:05    321#   $107
        INDIANAPOLIS-KANSAS CITY        385    2:52    300    100
                                       ------------------------------
             TOTAL                      818NM  5:57    621#   $207
                                       ==============================
```

```
                              ZULU        LOCAL
                              ----        -----
        DEPART WASHINGTON DULLES   1330     8:30 AM EST
        ARRIVE INDIANAPOLIS        1635    10:35 AM CST
        DEPART INDIANAPOLIS        1705    11:05 AM CST
        ARRIVE KANSAS CITY         1957     1:57 PM CST
```

Figure 25.6 The FAA-type flight plan for the third leg of the flight.

CHAPTER TWENTY-SIX

The G-WHIZ Computerised Flight Planning System

As you can tell by the S in "computerised," this program was developed in England and is marketed from there. The too-cute name "G-WHIZ" refers to the British equivalent to our N-numbers. The program runs on any 48K Apple computer with one or two disk drives (DOS 3.3), and requires a line printer capable of printing 132 characters per line.

The program includes a database consisting of some 230 European airports (aerodromes) and navigation aids. The database is capable of storing up to 700 destination coordinates, sorted alphabetically by the program. These are entered using latitude and longitude coordinates. You can erase the 230 European entries easily and replace them with others.

The program computes ground speed based on wind data, produces magnetic heading information including magnetic variations, computes fuel consumption, leg distances and time en route. It then produces a printed flight log that includes the results (see Figure 26.1).

The program includes an AUTOPLAN option that lets you save entire routes including an infinite number of waypoints for recall at any time. When using this option, the pilot need only key in wind and fuel-flow data. The program will read the file and compute headings, speed, time and fuel data and print the flight log.

The program is relatively fast. The entire database is read into memory whenever it is activated. The problem for use in the United States is that the user will have to create his or her own database, a time-consuming problem, but this needs to be done only once. A user-created database can be limited to the airports and nav aids located in the areas your flights are normally conducted. When additions are needed, they can always be added.

FOR CAPTAIN : BIGGLES   THE G-WHIZ COMPUTERISED FLIGHT PLANNING SYSTEM   DATE : 1/10/82

| FLIGHT LEVEL 100 | WIND 290/18 | MAGNETIC VARIATION + 7° W | TRUE AIR SPEED 135 | FUEL ON BOARD 56 | FUEL FLOW/HOUR 11.4 | ENDURANCE 4 H 54 M |

FROM	TO	FREQ	CO(M)	HDG(M)	G/S	LEG	FUEL	TIME	ETA	ATA
NOTTINGHAM	DAVENTRY VOR		189	* 197 *	137	44	4	19		
DAVENTRY VOR	BOVINGDON VOR		150	* 155 *	148	34	3	14		
BOVINGDON VOR	ELSTREE		124	* 126 *	152	9	1	4		
ELSTREE	LAMBOURNE VOR		099	* 098 *	153	18	1	7		
LAMBOURNE VOR	SOUTHEND		109	* 109 *	153	21	2	8		
SOUTHEND	DETLING VOR		200	* 208 *	134	16	1	7		
DETLING VOR	ASHFORD (TOWN)		141	* 145 *	150	14	1	6		
ASHFORD (TOWN)	LE TOUQUET		`150	* 155 *	148	46	4	19		
TOTALS						203 NM	15	1 H 24 M		

NOTE: NO LEGAL RESPONSIBILITY IMPLIED OR ACCEPTED. CHECK DATA PRIOR TO USE FOR NAVIGATIONAL PURPOSES.   (C) MFI 1982

Figure 26.1 A flight log prepared for a flight in England by the *G-WHIZ Computerized Flight Planning System.*

CHAPTER TWENTY-SEVEN

Software from PHH Aviation Systems

PHH Aviation Systems, Inc. of Golden, Colorado, is probably the leader in the area of sophisticated aviation software designed primarily for corporate flight departments operating one or more high-performance aircraft flown by professional crews. The family of programs marketed by PHH is the most complete I have ever seen. A brief description of these programs and their most important features follows.

AvPLANNER

This package of software programs helps plan fuel loads, compute direct routes for flights, and determine the best cruise altitude. The program produces a flight plan report that summarizes the navigation, performance and fuel requirements for a given flight. This information is computed from user entries outlining the flight, combined with information from aircraft, airport, nav aid and weather databases. The program is compatible with all weather database services that produce FAA format winds-aloft reports receivable as text files. It can be used with all types of aircraft, and permits user-customized databases for each aircraft. A major feature is the automatic design of area navigation routes. When departure and destination points have been entered, the program computes the location of evenly spaced waypoints, each linked with offset information to the nearest VORTAC or VOR/DME. The flight plan print-out (see Figure 27.1) lists the waypoints with an index number, longitude and latitude and offset data, all ready for input into the on-board RNAV equipment.

The program requires an IBM PC or XT with 192K RAM, IBM PC DOS 2.0 operating system, two disk drives, 80-column monitor, and an optional printer. Communication software is required if weather data are to be received from weather database services.

PERFORMANCE CHECKPOINTS:

Name	ETA	ATA	ETR H:MM	Alt MSL	Wind-Temp Frm/At+-C	TAS+WF =GS Kts Kts Kts	MAG Head	DIST NM	ETE H:MM
LAX/STARTUP	0:00	_____	2:27	126'		--Taxiing---		1.0	0:10
01 LAX/TAKEOFF	0:10	_____	2:17	136'	000/0+32	--Climbing--	302	34	0:06
02 FIM/VOR	0:16	_____	2:10	17446'	160/6-3	--Climbing--	339	96	0:14
03 Level off	0:30	_____	1:56	37000'	170/20-49	387+ 20=407	337	8.1	0:01
04 WAYPOINT #01	0:32	_____	1:55	37000'	186/19-49	386+ 19=405	337	104	0:15
05 WAYPOINT #02	0:47	_____	1:40	37000'	196/18-49	384+ 17=401	336	104	0:16
06 RNO/VOR	1:03	_____	1:24	37000'	220/18-49	382+ 10=392	326	110	0:17
07 WAYPOINT #03	1:20	_____	1:07	37000'	258/15-49	379- 1=378	325	110	0:18
08 WAYPOINT #04	1:37	_____	0:50	37000'	316/30-51	375- 27=348	324	110	0:19
09 WAYPOINT #05	1:56	_____	0:31	37000'	296/18-50	373- 12=361	322	91	0:15
10 Begin Descnt	2:11	_____	0:15	37000'	296/18-50	-Descending-	322	19	0:02
11 FINNY/INTX	2:14	_____	0:13	28123'	273/27-31	-Descending-	334	37	0:05
12 PRE-LAND FIX	2:19	_____	0:08	6935'	360/0+15	-Descending-	338	12	0:03
13 SEA/LANDING	2:22	_____	0:05	429'		--Taxiing---		.5	0:05
SEA/SHUTDOWN	2:27	_____	0:00	429'				0	0:00
TOTALS...		_____						839	2:27

FUEL CHECKPOINTS:

NAME	FUELFLOW LBS/HR	TO NEXT CHECKPT	SINCE STARTUP	FUEL ABOARD	FUEL TO SHUTDOWN	ACTUAL FUEL USED	ACTUAL FUEL ABOARD
LAX/STARTUP		100	0	5168	4043	_____	_____
01 LAX/TAKEOFF	Climb	511	100	5068	3943	_____	_____
02 FIM/VOR	Climb	701	611	4558	3433	_____	_____
03 Level off	1508	30	1311	3857	2732	_____	_____
04 WAYPOINT #01	1503	387	1342	3827	2702	_____	_____
05 WAYPOINT #02	1483	386	1729	3439	2314	_____	_____
06 RNO/VOR	1468	413	2115	3053	1928	_____	_____
07 WAYPOINT #03	1443	421	2528	2640	1515	_____	_____
08 WAYPOINT #04	1414	448	2949	2219	1094	_____	_____
09 WAYPOINT #05	1398	353	3397	1771	646	_____	_____
10 Begin Descnt	Descent	57	3750	1419	294	_____	_____
11 FINNY/INTX	Descent	139	3806	1362	237	_____	_____
12 PRE-LAND FIX	Descent	48	3945	1223	98	_____	_____
13 SEA/LANDING		50	3993	1175	50	_____	_____
SEA/SHUTDOWN		0	4043	1125	0	_____	_____

Average fuel consumption from takeoff to landing:
.215 nautical miles per pound.
1774.8 pounds per hour.

WEIGHT LIMITATIONS:

	WEIGHT IN POUNDS		
	MAXIMUM	PROJECTED	ACTUAL
Basic Operating Weight (includes crew)	21000	21000 ok	_____
+ Payload aboard	2480	1080 ok	_____
= Zero fuel weight	23480	22080 ok	_____
+ Usable fuel aboard	15514	5168 ok	_____
= Ramp gross weight	38800	27248 ok	_____
- Taxi to takeoff fuel used	15514	100 ok	_____
= Takeoff gross weight	38800	27148 ok	_____
- Fuel used enroute	15514	3893 ok	_____
= Destination landing gross weight....	35715	23255 ok	_____

Figure 27.1 A sample print-out produced by the AvPLANNER program.

AvNAV

This collection of 10 programs is an aid in flight planning. Each performs a specific flight-related calculation. The programs are:

DIST.EXE—the user enters two station identifiers, and the great circle distance between the two points is returned.

COURSE.EXE—when given two station identifiers, the great circle course outbound from the first identifier to the second is returned.

ETE.EXE—when provided with two station identifiers and estimated ground speed, returns the estimated time en route between the two points.

TAS.EXE—computes true airspeed from indicated airspeed, temperature and altitude data.

MACH.EXE—performs the same function, producing the Mach number instead of true airspeed.

DA.EXE—computes density altitude from pressure altitude and temperature data.

WINDTRI.EXE—asks that the pilot enter aircraft course, aircraft velocity, and wind direction and velocity. It computes the wind correction angle and ground speed.

SUN.EXE—uses date, airport identifier and time difference from Greenwich Mean Time to return time of morning twilight, morning sunrise, evening twilight and evening sunset.

UKNWIND.EXE—when given true airspeed, course and wind correction angle, computes wind direction and velocity.

XWIND.EXE—returns the crosswind component from course, wind direction and velocity data.

The program requires an IBM PC or XT with 192K RAM, the IBM-PC DOS 2.0 operating system and one disk drive.

AvMGR

AvMGR is more of a service providing computer-generated information rather than a program operated by the user. Aviation departments send

FLIGHT OPERATION SUMMARY
08/06/84 THRU 08/26/84

	THIS MONTH	***************** AVERAGE * PER MONTH	LAST TWELVE MONTHS HIGHEST MONTH	************* LOWEST * MONTH *	YEAR TO DATE
NUMBER OF FLIGHTS	40	47	79	21	89
POSITIONING FLIGHTS	10	6	17	1	10
AIRCRAFT HOURS FLOWN	45.6	67.0	99.6	35.9	135.6
POSITIONING HOURS	11.5	7.2	20.9	.5	11.5
AVERAGE FLIGHT DURATION	1.1	1.5	1.9	1.1	1.5
MILES FLOWN (STATUTE)	14779	23807	37827	12851	47792
PASSENGER MILES (STATUTE)	37016	66635	102834	30023	134404
PASSENGERS CARRIED	104	128	200	56	257
PASSENGERS PER FLIGHT	3.7	3.4	4.1	2.7	3.4
AIRCRAFT FLIGHT DAYS	18	16	20	9	46
AIRCRAFT USAGE CHARGES ($)	15080	1370	15080		15080

Figure 27.2 Flight operations summary sample print-out.

OPERATIONS OVERVIEW
08/06/84 THRU 08/26/84

RUN DATE: 08/28/84
RUN TIME: 09:35

A/C ID	FLT CAT	NO. FLTS	% FLTS	FLT HRS	BLK HRS	MILES FLOWN	MI /FLT	FLT SPD	BLK SPD	F HR /FLT	B HR /FLT	FUEL/ FLT/HR	FUEL BURNED	# OF PAX
N10P	L	9	39	11.7	12.7	5840	649	499	460	1.3	1.4	256	3000	20
	E	2	9	2.4	2.7	1120	560	467	415	1.2	1.4	167	400	6
	U	3	13	3.8	4.3	1590	530	418	370	1.3	1.4	197	750	10
	M	2	9	2.4	2.6	1180	590	492	454	1.2	1.3	292	700	4
	T	2	9	3.1	3.6	1000	500	323	278	1.6	1.8	139	430	8
	P	5	22	8.4	8.8	3890	778	463	442	1.7	1.8	224	1880	14
TOTAL		23	100	31.8	34.7	14620	636	460	421	1.4	1.5	225	7160	62

Figure 27.3 Operations overview sample print-out.

their flight logs, passenger manifests etc. to PHH in Golden, Colorado, and PHH produces monthly reports containing comprehensive summaries of the department's activities. Figures 27.2 through 27.10 are typical of the flight activity analyses produced by this service.

DISPATCH II and MANAGER I

These groups of programs run primarily on Data General hardware with RDOS, AOS or AOS/VS operating systems. If used with special interpretive software, they will also run on the IBM PC/XT and the AT.

AIRCRAFT PERFORMANCE ANALYSIS
08/06/84 THRU 08/26/84

RUN DATE: 08/28/84
RUN TIME: 09:38

A/C ID	DATE	ORIG	DEST	PIC	CAT	NO PAX	FLT HRS	BLK HRS	MILES FLOWN	PAX MILES	FLT SPD	BLK SPD	*FUEL PER* F/HR	*FUEL PER* B/HR	FUEL BURNED	MILES /LB
N10P	08/08	KDEN	KLAX	SSS	A	2	2.2	2.3	1210	2420	550	526	182	174	400	3.025
	08/08	KLAX	KSFO	SSS	B	2	1.1	1.2	620	1240	564	517	182	167	200	3.100
	08/08	KSFO	KDEN	SSS	Z	2	1.8	1.9	1100	2200	611	529	211	200	380	2.895
	08/09	KDEN	KKCK	NNN	D	4	1.3	1.5	500	2000	385	333	154	133	200	2.500
	08/09	KDEN	KKCK	NNN	D	4	1.3	1.5	500	2000	385	333	154	133	200	2.500
	08/09	KDEN	KKCK	NNN	D	4	1.3	1.5	500	2000	385	333	154	133	200	2.500
	08/09	KDEN	KKCK	NNN	D	4	1.3	1.5	500	2000	385	333	154	133	200	2.500
	08/09	KDEN	KKCK	NNN	D	4	1.3	1.5	500	2000	385	333	154	133	200	2.500
	08/09	KDEN	KKCK	NNN	D	4	1.3	1.5	500	2000	385	333	154	133	200	2.500
	08/09	KKCK	KDEN	NNN	E	4	1.8	2.1	500	2000	278	238	128	110	230	2.124
	08/10	KDEN	KSTL	BRT	A	2	1.0	1.1	590	1180	590	536	300	273	300	1.967
	08/10	KSTL	KDEN	BRT	A	2	1.2	1.3	590	1180	492	454	292	269	350	1.686
	08/10	KSTL	KDEN	BRT	A	2	1.2	1.3	590	1180	492	454	292	269	350	1.686
	08/10	KSTL	KDEN	BRT	A	2	1.2	1.3	590	1180	492	454	292	269	350	1.686
	08/10	KSTL	KDEN	BRT	A	2	1.2	1.3	590	1180	492	454	292	269	350	1.686
	08/10	KSTL	KDEN	BRT	A	2	1.2	1.3	590	1180	492	454	292	269	350	1.686
	08/10	KSTL	KDEN	BRT	A	2	1.2	1.3	590	1180	492	454	292	269	350	1.686
	08/10	KSTL	KDEN	BRT	A	2	1.2	1.3	590	1180	492	454	292	269	350	1.686
	08/10	KSTL	KDEN	BRT	A	2	1.2	1.3	590	1180	492	454	292	269	350	1.686
	08/10	KSTL	KDEN	BRT	A	2	1.2	1.3	590	1180	492	454	292	269	350	1.686
	08/12	KDEN	KQXI	BRT	1	2	2.8	3.0	1200	2400	429	462	268	288	750	1.600
TOTAL						23	62 31.8	35.1	14620	37240					7160	
AVERAGE							1.4	1.5	636		460	421	225	206		2.042

Figure 27.4 Aircraft performance analysis sample print-out.

AIRCRAFT UTILIZATION BY AUTHORIZER
08/13/84 THRU 09/26/84

RUN DATE: 09/27/84
RUN TIME: 13:45

AUTH	A/C ID	DATE	CAT	FROM	TO	NO. PAX	MILES FLOWN	PAX MILES	FLT HRS	BLK HRS	$$$$$$	NO. FLTS
	N10	09/17/84	E	KDEN	KSTL	4	676	2704	1.7	1.9	0	
		TOTAL				4	676	2704	1.7	1.9	0	1
	TOTAL					4	676	2704	1.7	1.9	0	1
0001	N10	09/14/84	E	KDEN	KGRI	4	309	1236	1.0	1.6	800	
		09/14/84	E	KGRI	KOKC	4	336	1344	1.2	1.3	900	
		09/14/84	C	KOKC	KORD		601		1.5	2.1	1300	
		09/14/84	C	KORD	KDEN		781		2.3	2.1	2000	
		09/15/84	T	KDEN	KDAL	1	568	568	1.3	1.8	1300	
		09/15/84	M	KDAL	KOKC	1	157	157	.6	1.1	1350	
		09/15/84	E	KOKC	KHOU	3	364	1092	1.3	1.3	1000	
		09/15/84	E	KHOU	KDEN	4	769	3076	2.3	3.1	2020	
		09/16/84	P	KDEN	KSTL	2	676	1352	1.5	2.1	1300	
		09/16/84	F	KSTL	KPOU	4	774	3096	1.8	2.0	2200	
		09/16/84	G	KPOU	KHPN	4	35	140	.2	.8	250	
		09/16/84	E	KHPN	KORD	4	639	2556	2.0	2.6	1750	
		09/16/84	P	KORD	KDEN		781		2.3	2.8	2000	
		TOTAL				31	6790	14617	19.3	25.7	18170	13
0001	N162A	08/13/84	E	KDEN	KHPN	3	1418	4254	3.6	3.8	3610	
		08/14/84	E	KHPN	KDEN	7	1418	9926	4.5	4.6	4445	
		08/15/84	E	KDEN	KGRI	8	309	2472	1.0	1.1	1033	
		08/15/84	E	KGRI	KTUL	8	308	2464	1.1	1.2	1130	
		08/15/84	E	KTUL	KORD	8	508	4064	1.7	1.8	1770	
		08/15/84	E	KORD	KDEN	8	781	6248	2.5	2.7	2600	
		TOTAL				42	4742	29428	14.4	15.2	14588	6
0001	TOTAL					73	11532	44045	33.7	40.9	32758	19
021	N10	08/13/84	E	KDEN	KSTL	4	676	2704	1.7	1.9	1300	
		08/13/84	E	KSTL	KDEN	3	676	2028	.8	.9	850	
		TOTAL				7	1352	4732	2.5	2.8	2150	2
021	TOTAL					7	1352	4732	2.5	2.8	2150	2

Figure 27.5 Aircraft utilization by authorizer analysis sample print-out.

PASSENGER OPERATIONS ANALYSIS
08/06/84 THRU 08/26/84

RUN DATE: 08/28/84
RUN TIME: 09:40

A/C ID	AUTH DEPT	NO. PAX	% TOT	NO. FLTS	% TOT	PAX MILES	% TOT	SEAT MILES	% TOT	AVG PAX PER FLT	$$$$$$	% TOT
N10		4	2.8	1	2.1	2704	3.2	2704	2.5	4.0		
	0001	31	21.4	13	27.7	14617	17.4	27160	25.0	2.4	18170	28
	021	7	4.8	2	4.3	4732	5.6	5408	5.0	3.5	2150	3
	1	13	9.0	5	10.6	7663	9.1	10400	9.6	2.6	6966	10
	10	3	2.1	1	2.1	2523	3.0	3364	3.1	3.0	1575	2
	3	37	25.5	11	23.4	23423	27.8	28172	25.9	3.4	17089	26
	4	4	2.8	1	2.1	3364	4.0	3364	3.1	4.0	1645	2
	L113	6	4.1	2	4.3	1344	1.6	1792	1.6	3.0	1620	2
	ROBT	40	27.6	11	23.4	27365	28.2	26304	24.2	3.6	15758	24
	TOTAL	145	100.0	47	100.0	84135	100.0	108668	100.0	3.1	64973	100
N162A	0001	42	12.9	6	10.5	29428	12.4	37936	11.5	7.0	14588	11
	2244	20	6.2	6	10.5	34197	14.5	70688	21.4	3.3	25295	19
	23	16	4.9	2	3.5	21664	9.2	21664	6.6	8.0	7710	6
	3	179	55.1	26	45.6	117463	49.7	131792	39.9	6.9	52295	41
	444	24	7.4	3	5.3	16200	6.8	16200	4.9	8.0	6190	4
	ROBT	44	13.5	14	24.6	17604	7.4	52368	15.8	3.1	20900	16
	TOTAL	325	100.0	57	100.0	236556	100.0	330648	100.0	5.7	126978	100
N700	055	47	29.6	13	35.1	24952	20.5	58272	26.9	3.6	22335	28
	OMM	10	6.3	3	8.1	8853	7.3	21400	9.9	3.3	8110	10
	1	14	8.8	4	10.8	11215	9.2	25696	11.9	3.5	10070	12
	2	20	12.6	6	16.2	12774	10.5	26928	12.4	3.3	10785	13
	ANDY	52	32.7	7	18.9	45160	37.2	47432	21.9	7.4	15028	19
	ROBT	16	10.1	4	10.8	18528	15.3	37056	17.1	4.0	12650	16
	TOTAL	159	100.0	37	100.0	121482	100.0	216784	100.0	4.3	78978	100
FLEET	TOTAL	629		141		442173		656100		4.5	270929	

Figure 27.6 Passenger operations analysis sample print-out.

The total program package may be thought of as a sort of "super clerk," gathering all the latest information on all major functions in the flight department. It correlates the data and gives them back to the user upon command in visual displays or as printed copy. It can produce just a few key facts on which to base an immediate decision, or provide detailed historical reports on various aspects of the operation for planning or budgeting.

A few keystrokes provide all information needed to schedule aircraft, passengers and crews. The system displays or prints flight itineraries, passenger manifests, wait-listed passengers and pertinent messages, and produces complete passenger travel itineraries. It generates crew schedules showing currency requirements, time off, training, and special scheduled activities. It displays flight schedules, lists complete aircraft itineraries and automatically computes segment mileage, time en route, ETD or ETA.

A stand-alone module, called *Management Information System*, generates operational reports covering any selected time span on any aircraft or group of aircraft. These reports can be used to improve aircraft use through analysis of factors such as flight stage length and hours per flight, aircraft performance, the use of aircraft by company departments, load factors, profiles of missions flown, and fuel management.

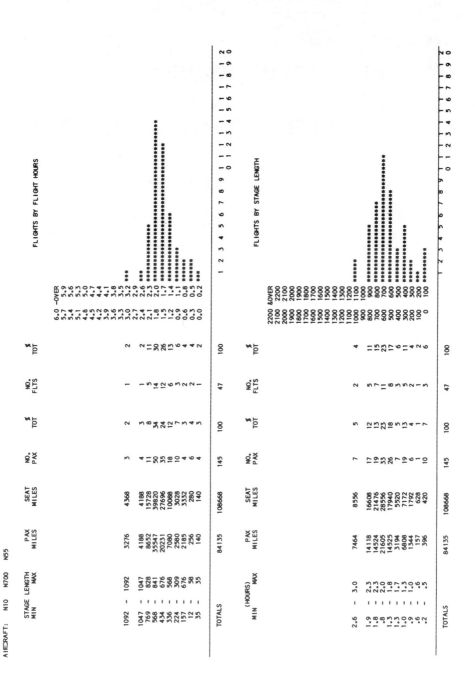

Figure 27.7 Mission profile analysis sample print-out.

AIRPORT PAIR ANALYSIS
08/06/84 THRU 08/26/84

RUN DATE: 08/28/84
RUN TIME: 09:50

AIRCRAFT: N10   N700   N162A   N55

FROM	TO	NO. FLT	NO. PAX	AVG PAX	AVG DIST	TTL HRS	(AVERAGE HRS) FLT	BLK	********* CONSUMED	FUEL ********* AVG BURN	NW/ UNIT	TOTAL $$$$$$$
KATL	KMIA	1	4	4.0	517	1.5	1.5	1.7	1530	1530	.338	1150
KATL	KBNA	1	8	8.0	185	.6	.6	.8	1200	1200	.154	750
MXKF	KNEW	1	4	4.0	1305	3.4	3.4	3.6	4200	4200	.311	2830
KBIL	KDEN	1	1	1.0	396	1.3	1.3	1.4	2000	2000	.198	1340
KBIL	KSEA	2	11	5.5	576	3.3	1.7	1.7	4500	2250	.256	3233
KBOS	KDCA	1	3	3.0	345	1.4	1.4	1.5	2000	2000	.173	1430
KORD	KDEN	5	16	3.2	781	11.9	2.4	2.9	14700	2940	.266	11270
KORD	KMSP	1	8	8.0	289	1.2	1.2	1.4	1900	1900	.152	1275
KORD	KJFK	1	3	3.0	641	1.9	1.9	2.1	1940	1940	.330	1570
KORD	KSTL	1	2	2.0	224	.9	.9	1.2	1000	1000	.224	820
KDAL	KDEN	2	8	4.0	568	3.8	1.9	2.2	4975	2488	.228	1763
KDAL	KHOU	1	5	5.0	208	1.0	1.0	1.1	1500	1500	.159	1000
KDAL	KJFK	1	8	8.0	1198	3.2	3.2	3.3	4400	4400	.272	3175
KDAL	KOKC	1	1	1.0	157	.6	.6	1.1	2500	2500	.063	1350
KAPA	KSFO	2	3	3.0	1279	3.9	3.9	4.1	5300	5300	.241	4010
KAPA	KGRI	1	6	6.0	311	1.8	.9	1.3	1500	750	.415	1990
KDEN	KTUL	2	2	2.0	470	1.3	1.3	1.5	1500	1500	.313	1300
KDEN	KATL	1	12	6.0	1047	5.4	2.7	2.9	6500	3250	.322	4850
KDEN	KBIL	1	4	4.0	396	1.2	1.2	1.3	990	990	.400	850
KDEN	KORD	3	8	8.0	781	2.3	2.3	2.5	3500	3500	.223	2440
KDEN	KDAL	2	4	1.3	568	4.4	1.5	1.8	5450	1817	.313	4330
KDEN	KAPA	2	7	3.5	12	.5	.3	.3	1400	700	.017	1200
KDEN	KDTW	1	0	.0	1028	2.5	2.5	2.9	2400	2400	.428	2200
KDEN	KGRI	2	12	6.0	309	2.0	1.0	1.4	2500	1250	.247	1833
KDEN	KMIA	2	12	6.0	1488	8.4	4.2	4.3	10600	5300	.281	8240
KDEN	KMKE	3	18	6.0	787	6.4	2.1	2.4	8400	2800	.281	5905
KDEN	KMSP	2	2	1.0	601	3.1	1.6	1.9	3600	1800	.334	2550
KDEN	KMRY	1	3	3.0	821	2.0	2.0	2.2	2150	2150	.382	1633
KDEN	KJFK	1	4	4.0	1418	3.6	3.6	3.8	4700	4700	.302	5650
KDEN	KORF	1	8	8.0	1354	3.7	3.7	3.8	5100	5100	.265	3800
KDEN	KOKC	1	3	3.0	434	1.3	1.3	1.5	2100	2100	.207	1330
KDEN	KSAN	1	8	8.0	728	2.3	2.3	2.3	3300	3300	.221	2335
KDEN	KSFO	1	0	.0	828	2.8	2.8	3.1	4100	4100	.202	3300
KDEN	KSEA	1	8	8.0	884	2.7	2.7	2.8	4000	4000	.221	2850
KDEN	KSTL	4	13	3.3	676	6.7	1.7	2.0	6850	1713	.395	4210

Figure 27.8 Airport pair analysis sample print-out.

CREW LOG REPORT
08/02/84 THRU 08/16/84

RUN DATE: 08/27/84
RUN TIME: 14:57

CREW # 01 NAME LEWIS

DATE	AIRCRAFT TYPE	ID	C A T	ORIG	DEST	DIST	NO. PAX	FLT HRS	BLK HRS	PIC HRS	SIC HRS	*LANDINGS* DAY	NITE	INSTRUMENT HRS	INSTRUMENT APRCHS	X-C HRS	SIMULATOR HRS
08/02	DA10	N10	P	KTEB	KEYN	60	0	.5	.7	.5		1		.2	1	.5	
08/02	DA10	N10	E	KEYN	KEVT	220	4	.9	1.1		.9					.9	
08/03	DA10	N10	C	KEVT	KEYN	220	2	.7	.8		.7					.7	
08/04	DA10	N10	E	KEYN	KEVT	235	3	1.0	1.2	1.0		1		.3	1	1.0	
08/04	DA10	N10	M	KEVT	KEVT	20	0	.3	.7	.3		1		.1	1	.3	
08/04	DA10	N10	E	KEVT	KEYN	220	2	.7	.8	.7			1		3	.7	
08/04	DA10	S												3.0	1		3.0
08/06	DA10	N10	C	KEYN	KEVT	235	1	1.0	1.2	1.0		1		.3	1	1.0	
08/07	DA10	N10	C	KEVT	KEYN	225	1	.8	1.0	.8		1		.2	1	.8	
08/08	DA10	N10	P	KEYN	KECT	220	0	.9	1.1	.9		1			1	.9	
08/08	DA10	N10	E	KECT	KJFK	240	5	1.0	1.5		1.0					1.0	
08/08	DA10	N10	P	KJFK	KEVT	240	0	1.1	1.2	1.1			1	.3	1	1.1	
08/09	DA10	N10	E	KEVT	KHVL	240	3	1.1	1.2		1.1					1.1	
08/09	DA10	N10	E	KHVL	KEVT	240	3	1.1	1.2	1.1		1		.7	1	1.1	
08/09	DA10	N10	P	KEVT	KEYN	245	0	1.1	1.2		1.1					1.1	
08/12	DA10	N10	E	KEYN	KMIA	1026	4	2.9	3.2		2.9					2.9	
08/12	DA10	N10	P	KMIA	KEVT	1230	1	2.9	3.0	2.9		1				2.9	
08/12	DA10	N10	E	KEVT	KEYN	225	5	.8	.8	.8		1		.2	1	.8	
08/13	DA10	N10	E	KEYN	KEVT	228	2	.9	1.0		.9					.9	
08/13	DA10	N10	E	KEVT	KDAL	1227	5	3.0	3.3	3.0		1		.6	1	3.0	
08/13	DA10	N10	E	KDAL	KMSY	379	5	1.3	1.4		1.3					1.3	
08/16	DA10	N10	E	KMSY	KEVT	816	5	2.4	2.4	2.4		1		.8	1	2.4	
THIS MONTH: DA10	TOTALS						49	26.4	30.0	16.5	9.9	11	2	3.7	16	26.4	
SIMULATOR	TOTALS													3.0			3.0
GRAND	TOTALS						49	26.4	30.0	16.5	9.9	11	2	6.7	16	26.4	3.0

Figure 27.9 Crew log report sample print-out.

Another module, *Flight Plan Interface*, determines fuel consumption under different wind, temperature, climb and cruise level conditions. It computes the route that results in minimum time track on the airway system, on canned routes or random tracks. It includes airways in the flight plan where applicable, along with SIDs and STARs, and it generates a second computed flight plan to the alternate, including all the features of the primary flight plan.

The *Commercial Travel Interface* module calls up all company personnel booked on commercial flights, along with other travel information useful to improving load factors of company aircraft.

The *CAMP/Dispatch* maintenance module automates and displays all aircraft maintenance records, and shows all inspections, services, and component replacement requirements for a selected period. The system compiles a cumulative history of all scheduled and unscheduled maintenance performed on each aircraft.

The *Inventory Control* module provides an instant summary of all parts on hand, on order and overdue from suppliers. In addition, it compiles a repair history, procurement history and component history.

```
Editing Record 2 of 2                                                    FORM
Enter Start Date in the form: MM/DD/YY

     Start Date:    11/02/84   Trip ID:  BOS-LGA/1002___   Aircraft: N10P
     Tail #:        N10P__     Manifest: Tim Johnson, Fr   Res#: BOS-LGA/1002
     Leg#:          1_                                     Date: 11/2/84
     From:          KBOS       To:          KLGA           Leg#: 1
     Dist:          284_       Fuel Used:   2010__                 MANIFEST
     Block Time:    1.2        Flight Time: 1__            1: Tim Johnson
     Pax:           0_         Auth:        LEWIS/12        2: Frank Lewis
     Cat:           P                                      3: Allan Webber
                                                           4: Jean Andrews
     Capt:          KES                                    5: Bob Thompson
     Capt PIC HRS:  1__    Capt SIC HRS:   0__             6: Rita Jordan
     Capt DAY LND:  1      Capt NIGHT LND: 0               7:
     Capt NIGHT HRS: 0__   Capt X-C HRS:   1__             8:
     Capt ACT INST: 0.2    Capt INST APP:  1               9:
                                                           10:
     F/O:           RDL                                    11:
     F/O PIC HRS:   0__    F/O SIC HRS:    1__             12:
     F/O DAY LND:   1      F/O NIGHT LND:  0               13:
     F/O NIGHT HRS: 0__    F/O X-C HRS:    1__             14:
     F/O ACT INST:  0.2    F/O INST APP:   1               15:
                                            DATA_ENTRY               MANIFEST
     03-Dec-84  04:40 PM                               Calc



                      Passenger Manifest Report      Run Date: 12/03/84
                        8/4/84 Thru 10/18/84          Run Time: 16:35:45


        *** Flight Information ***          *** Passenger Names ***
      Trip Start Date = 11/2/84           1: Jim Reed
              Trip ID = XT-3414/R         2: Allan R. Bobbe
           Leg Number = 1                 3: Frank Johnson
             Aircraft = N10P              4: Bob Lewis
                 From = KDEN
                   To = KHNL
              Captain = KES
        First Officer = 6
           Authorizor = REED
             Category = P
             Leg Cost = $2795
       Cost/Passenger = $699
```

Figure 27.10 Passenger manifest report sample print-out.

Manager I is intended primarily for fixed base operators and air taxi
operators. It provides an orderly flow of detailed operational and financial information for business management and planning. Interacting with
it is the *Aviation Management Work Order System* that offers on-line
control of maintenance shop operations. It calculates and prints invoices,
tracks status of work orders, gathers data needed for evaluating flat-rate
fees, and helps to monitor the performance of maintenance personnel.

CHAPTER TWENTY-EIGHT

The Dow-4 Gazelle

The Dow-4 Gazelle is a program that simulates a typical four-seat, single-engine, instrument-equipped, piston engine aircraft. It can be used to learn to fly from scratch, or to practice ILS approaches and other instrument procedures. The program is written in Texas Instruments BASIC, and runs on the TI-99/4 computer equipped with a joystick. It is available on cassette or disk.

The instrument panel, occupying most of the display, consists of an artificial horizon, directional gyro, airspeed indicator, altimeter, turn and bank and vertical speed indicators, navigation radio with OBI and ILS displays, marker beacon lights, fuel quantity gauges for two tanks, tachometer, flap setting indicator, wing leveler and stall warning light. The joystick is used to control pitch and bank, and the keyboard is employed to effect changes in power, raise or lower flaps, turn the wing leveler on or off, etc.

The dials and lights on the instrumental panel respond in less than a second on the average, meaning that the user must beware not to over-control when, initially, nothing happens after a control input. The program includes sound effects, such as engine noise that varies with power and airspeed, the stall warning horn and the sounds associated with inevitable crashes.

The manual that comes with the program, though less than impressive in appearance (reproduced in typewriter type), is well written and consists of clear instructions. It also includes a separate section designed for the novice pilot, entitled "The Basics of Flying."

CHAPTER TWENTY-NINE

Compuflight
Operations Service

This is a computerized flight-planning service designed for the "bigger boys," meaning the large corporate flight operations and air charter operators as well as scheduled and unscheduled passenger and freight carriers.

Headquartered in Fort Washington on Long Island, close to John F. Kennedy and LaGuardia airports, Compuflight uses a bank of Digital PDP-11 series computers that contain proprietary software, enabling users to immediately plan flights worldwide. The databases include specifications and performance parameters for over 80 kinds of aircraft, as well as current information about airports, navigation aids, established air routes and alternate routes worldwide (except for the East Block countries). Individual aircraft registration numbers are stored for all customers, and fixed routes are stored where required by FAA and/or when requested by the customer.

Customers can access the databases in any of several ways: direct via dedicated terminals in their own offices, via the worldwide ARINC or SITA networks, or through direct-dial connections using either TYMNET or the telephone.

Prestored routes are built to ATC and customer requirements, and include SIDs and STARs. Flight plans are filed automatically with domestic or foreign air-traffic control authorities, meeting all FAA, ATC and ICAO regulations. The service includes simplified re-clear or re-release procedures to increase payload and/or to reduce fuel consumption in accordance with FAA requirements. A minimum time track program can be used to build minimum time routes or it can steer the aircraft to a fix or fixes to avoid weather.

The service provides immediate access to weather data from the U.S. Weather Service 24 hours a day, seven days a week, via data link with WSI or Global Weather Dynamics.

Compuflight was founded by two former airline executives in the wake of the recent airline deregulation, which has created large numbers of companies that do not have the in-house computerized flight-planning capabilities owned and operated by the major air carriers. The company's success in this venture is demonstrated by the fact that it serves over 40 customers, including People Express, Federal Express, Emery Worldwide, United Parcel, Purolator, Regent Air and American Trans Air.

DO-IT-YOURSELF
AVIATION
PROGRAMS

Do-It-Yourself Programs

In this section of the book, we'll look at a library of aviation programs you can key into your own computer (or you can purchase them from the publisher along with some general-purpose utility programs not included here). The programs are written in BASIC-80 on an Apple II Plus equipped with Z-80-based Microsoft CP/M. They will run without changes or modification on any Apple II, II Plus or IIe comparably equipped. For use with other computers or Apples without CP/M, certain changes will have to be made. Most modifications are described in detail in Appendix A.

No programming experience is required to type the programs into your own computer. Some familiarity with programming techniques may be helpful, however, because it is difficult to avoid minor typing errors that can cause the program to refuse to function.

To simplify the process of finding and correcting errors—debugging the program—the text accompanying each program includes brief descriptions of what takes place in its different sections. These should be used as aids, suggesting when to perform intermediate runs to make sure the program functions properly up to that given point.

To use these programs you need a computer equipped with at least one disk drive. A line printer, while not absolutely necessary, would be most helpful. None of the programs exceed 15k bytes in length, but if they are used as written (in BASIC-80) the random access memory (RAM) of your computer must at least be 48K. MBASIC and the CP/M system, which must be on the disk in order for it to boot and run, occupy approximately 30K.

To prepare your disk for program writing in BASIC-80, you must perform a number of steps that involve several of the programs that reside on your CP/M disk. Assuming that you're operating with two disk drives, the steps are:

1. Insert the CP/M disk into Drive A and your work disk into Drive B and boot the CP/M disk.

2. Type FORMAT B : and press the RETURN key to format your work disk.

3. When the formatting process is completed, type COPY B : =A : /S and press the RETURN key to copy the CP/M system onto your work disk.

4. Next type P I P B : MB . COM = A : MBAS I C . COM and press the RETURN key which will cause MBASIC to be copied onto your work disk under the abbreviated file name MB.COM.

You're now ready to key in programs. There are two methods you can use. If you use the EDIT function available with BASIC-80 (MBASIC), boot your work disk and wait for the A > prompt to appear (assuming the disk is in Drive A). Now type MB and press the RETURN key and the A > prompt will be replaced by the BASIC-80 prompt: Ok. You can now type the program and save it on your work disk. The command to use is SAVE "F I LENAME . EXT" or SAVE "F I LENAME . EXT" , A, replacing FIL-ENAME with any file name of eight or less characters. .EXT should be any extension of up to three characters. If the first SAVE version is used, the program will be recorded in binary format. The second version stores it in ASCII format, which takes up more space on the disk.

The second method, most preferable, is to use a word processor for program entry. All programs reproduced here were written using *WordStar*, and that is the process I'll describe. Other word processors may require different steps.

Insert the *WordStar* disk into Drive A and the work disk into drive B. Change the logged disk to the B drive. With the NO-FILE MENU in display, type N to create a non-document file— a program. At this point type your program, using all the editing functions available in your word processor. It is always a good idea to save your work frequently (type CTRL K S) to avoid losing it through an inadvertent power failure or some other mishap. When you're ready to run all or part of the program, type CTRL K X to return to the CP/M system. The B > prompt will appear in the lower left-hand corner of your screen. Now type MB F I LENAME . EXT and press the RETURN key. After a second or two your program will run. You can list the program by typing L I ST in reply to Ok prompt, or send it to the line printer by typing LL I ST.

To make corrections, either simply retype the to-be-corrected program lines and then run the program again (simply type RUN and press the RETURN key) to make sure the corrections you made are functioning properly. You can return to *WordStar* by typing SYSTEM and pressing the RETURN key; the OK prompt will be replaced by the B> prompt. Then type A : WS and press the RETURN key; your display will be replaced by the NO-FILE MENU. Return to your program listing by typing N. Make the necessary corrections and then repeat the process.

If you opt to make corrections while in BASIC-80, there are some things to remember. If you've made your corrections satisfactorily and then return to *WordStar*, you'll find that what appears on the screen is the uncorrected version, because you've failed to save the corrections. To have *WordStar* display the corrected version, first type SAVE "F I LE-NAME . EXT", A and press the RETURN key. Be sure to include , A to save your work in the ASCII format. A program saved in the binary format and then displayed under *WordStar* will consist of some totally unintelligable garbage. If you accidentally save it without the ,A command, all is not lost. *WordStar* (and many other word processors) automatically creates a back-up file called FILENAME.BAK that contains the previous version of the program. To access it, rename it, (type E and enter the file name with an extension other than .BAK). If you want to use the same extension you used before, first delete the binary program saved using that extension, because you can't have two files with the same file name-extension combination on the same disk.

Many of the programs included here create their own text files that are automatically stored on the disk containing the main program. For that reason it's a good idea to make sure the unused space on the disk is sufficient. This is especially important if you intend to use the *Flight Route File* program to store many route segments.

CHAPTER THIRTY-ONE

Weight and Balance for Any Aircraft

For some reason, most pilots have considerable difficulty determining the weight-and-balance condition of their aircraft based on the passenger, baggage and/or freight load for a given flight. This program handles that problem with ease, but requires a large initial input to customize it for the aircraft with which it will be used.

The program can accommodate aircraft with up to ten seats (five rows of two seats each), and can be used just as effectively with two-, four-, and six-seat aircraft. For these planes, the program lines for non-existing seat rows may be eliminated. If the aircraft does not have a nose baggage compartment, eliminate line 530, and so on.

The main menu offers the choice of using previously-entered aircraft weight and moment data, entering a new set of such data, or looking at data stored earlier:

```
Main menu:
---------------------------------------------------------------
1    Use previously entered data.
2    Enter new weight and moment data.
---------------------------------------------------------------
3    Review previously entered data.
---------------------------------------------------------------
4    Exit program.
---------------------------------------------------------------
Which?
```

To customize the program for your aircraft, select the second choice. This produces prompts for the data to be entered and saved for future use:

```
Aircraft identification (to 28 letters)
N-Number?
```

```
Basic operating weight?
Zero fuel weight?
Ramp weight?
Takeoff weight?
Landing weight?
```

```
Empty aircraft moment?
Seat row 1 moment?
Seat row 2 moment?
Seat row 3 moment?
Seat row 4 moment?
Seat row 5 moment?
Nose baggage compartment moment?
Aft cabin baggage area moment?
Rear baggage bay moment?
Fuel moment?
Forward CG limit?
Aft CG limit?
```

All these data should be entered as the maximum allowable values listed in the aircraft's owner's manual. The moment data are moment arms. Luckily, all this needs to be keyed in only once.

At some later date when equipment changes or some other reason requires a change in any of the values, you may review the previously entered data by selecting the third choice from the main menu. Then you can switch to the second choice and change data selectively as needed.

The first choice from the menu is the one you'll use most of the time. It asks that the weight data for the current flight be entered as follows:

```
Key in data in pounds. For empty type >0<
```

```
Pilot weight?
Copilot weight?
Number of passenger seats used?
Row 2, left seat passenger weight?
Row 2, right seat passenger weight?
```

and so on, for as many seat rows as are being used. Then it continues:

```
Nose compartment baggage weight?
Aft cabin area baggabe weight?
Rear bay baggage weight?
```

The program uses these figures to produce a display of total weights up to this point:

```
Total seat weight:          xxxx pounds.
Total baggage weight:       xxxx pounds.
Payload:                    xxxx pounds.
----------------------------------------------------------------------------
Zero fuel weight:           xxxx pounds. O.K.
Fuel on board (pounds)?
```

After the fuel on board is keyed in, the program computes the ramp weight, take-off and landing weights of the aircraft:

```
Ramp weight:                xxxx pounds. O.K.
Fuel used for taxi and holding?
Takeoff weight:             xxxx pounds. O.K.

Estimated fuel burn during trip?
Estimated landing weight:   xxxx pounds. O.K.
```

If the weight is not within the allowable limits, instead of displaying "O.K." the program displays any of these lines:

```
(weight figure) is above the max zero fuel weight of xxxx pounds.
(weight figure) is above the max ramp weight of xxxx pounds.
(weight figure) is above the max take-off weight of xxxx pounds.
(weight figure) is above the max landing weight of xxxx pounds.
```

Immediately after a satisfactory take-off weight has been displayed, the program responds with one of the following, depending on the CG condition:

```
The CG is xxx and is within acceptable limits.
The CG is xxx and is beyond the forward limit.
The CG is xxx and is beyond the aft limit
Weight must be redistributed!
```

Lines 100–290 display the menu. Lines 310–440 accept the weight data and store them in a file called WEIGHT.DTA. Lines 450–640 store the moment data in a file called MOMENT.DTA. Lines 650– 910 access the two files and display the previously entered data. Lines 920 through the end are used to key in the weight data for a given flight, perform the necessary computations, and display the results (see Figure 31.1A–C).

```
100 REM WEIGHT AND BALANCE FOR ANY AIRCRAFT
110 REM BASIC-80
120 GOTO 190
130 PRINT STRING$(80,45):RETURN
140 HOME:VTAB(10):RETURN
150 PRINT:INPUT "Press >RETURN< (Q to quit)  ",R$:IF R$="Q" THEN 160 ELSE RETURN
160 GOSUB 140:GOSUB 130:PRINT TAB(38)"End.":GOSUB 130:END
170 PRINT:INPUT "Entries correct?  (Y/N)  ",CORRECT$:RETURN
180 PRINT:INPUT "Entry correct?  (Y/N)  ",CORRECT$:RETURN
190 GOSUB 140:GOSUB 130
200 PRINT TAB(10)"This program determines the weight and balance condition for any":PRINT
210 PRINT TAB(10)"aircraft with up to 10 seats for which the appropriate data have":PRINT
220 PRINT TAB(10)"previously been entered.":GOSUB 130
230 GOSUB 150:GOSUB 140
240 PRINT TAB(10)"Main menu:":GOSUB 130
250 PRINT 1,"Use previously entered data."
260 PRINT 2,"Enter new weight and moment data.":GOSUB 130
270 PRINT 3,"Review previously entered data.":GOSUB 130
280 PRINT 4,"Exit program.":GOSUB 130
290 INPUT "Which?  ",WHICH:GOSUB 140
300 ON WHICH GOTO 920,310,650,160
310 OPEN "R",#1,"WEIGHT.DTA",47
320 FIELD #1,20 AS T$,6 AS U$,4 AS V$,4 AS W$,4 AS X$,4 AS Y$,4 AS Z$
330 INPUT "Aircraft identification (20 letters) ",ACR$
340 INPUT "N-number?                            ",NUM$
350 INPUT "Basic operating weight?              ",BOW
360 INPUT "Zero fuel weight?                    ",ZFW
370 INPUT "Ramp weight?                         ",RW
380 INPUT "Takeoff weight?                      ",TOW
390 INPUT "Landing weight?                      ",LDW
400 GOSUB 170:IF CORRECT$<>"N" THEN 410 ELSE GOSUB 140:GOTO 330
410 LSET V$=MKI$(BOW):LSET W$=MKI$(ZFW):LSET X$=MKI$(RW):LSET Y$=MKI$(TOW):LSET Z$=MKI$
(LDW)
420 LSET T$=ACR$:LSET U$=NUM$
430 PUT #1,1
440 CLOSE #1
450 OPEN "R",#2,"MOMENT.DTA",46
460 FIELD #2,6 AS A$,3 AS B$,3 AS C$,3 AS D$,3 AS E$,3 AS F$,3 AS G$,3 AS H$,3 AS I$,5 A
S J$,5 AS K$,5 AS L$
470 INPUT "Empty aircraft moment?               ",EAM
480 INPUT "Seat row 1 moment?                   ",SR1
490 INPUT "Seat row 2 moment?                   ",SR2
500 INPUT "Seat row 3 moment?                   ",SR3
510 INPUT "Seat row 4 moment?                   ",SR4
520 INPUT "Seat row 5 moment?                   ",SR5
530 INPUT "Nose baggage compartment moment?     ",NBC
540 INPUT "Aft cabin baggage area moment?       ",ACB
550 INPUT "Rear baggage bay moment?             ",RBB
560 INPUT "Fuel moment?                         ",FUM
570 INPUT "Forward CG limit?                    ",FCG
580 INPUT "Aft CG limit?                        ",ACG
590 GOSUB 170:IF CORRECT$<>"N" THEN 600 ELSE GOSUB 140:GOTO 470
```

```
600 LSET A$=MKI$(EAM):LSET B$=MKI$(SR1):LSET C$=MKI$(SR2):LSET D$=MKI$(SR3)
610 LSET E$=MKI$(SR4):LSET F$=MKI$(SR5):LSET G$=MKI$(NBC):LSET H$=MKI$(ACB)
620 LSET I$=MKI$(RBB):LSET J$=MKI$(FUM):LSET K$=MKI$(FCG):LSET L$=MKI$(ACG)
630 PUT #2,1
640 CLOSE #2:GOSUB 130:GOSUB 150:GOSUB 140:GOTO 240
650 OPEN "R",#1,"WEIGHT.DTA",47
660 FIELD #1,20 AS T$,6 AS U$,4 AS V$,4 AS W$,4 AS X$,4 AS Y$,4 AS Z$
670 GET #1,1
680 PRINT"Aircraft identification:              ";T$
690 PRINT"N-number:                            ",U$:GOSUB 130
700 PRINT"Basic operating weight:              ";CVI(V$)
710 PRINT"Zero fuel weight:                    ";CVI(W$)
720 PRINT"Ramp weight:                         ";CVI(X$)
730 PRINT"Takeoff weight:                      ";CVI(Y$)
740 PRINT"Landing weight:                      ";CVI(Z$):GOSUB 130
750 CLOSE #1:GOSUB 150:GOSUB 140
760 OPEN "R",#2,"MOMENT.DTA",46
770 FIELD #2,6 AS A$,3 AS B$,3 AS C$,3 AS D$,3 AS E$,3 AS F$,3 AS G$,3 AS H$,3 AS I$,5 AS
J$,5 AS K$,5 AS L$
780 GET #2,1
790 PRINT"Empty aircraft moment:               ";CVI(A$)*1000
800 PRINT"Seat row 1 moment:                   ",CVI(B$)
810 PRINT"Seat row 2 moment:                   ",CVI(C$)
820 PRINT"Seat row 3 moment:                   ",CVI(D$)
830 PRINT"Seat row 4 moment:                   ",CVI(E$)
840 PRINT"Seat row 5 moment:                   ",CVI(F$):GOSUB 130
850 PRINT"Nose baggage compartment moment:     ",CVI(G$)
860 PRINT"Aft cabin baggage area moment:       ",CVI(H$)
870 PRINT"Rear baggage bay moment:             ",CVI(I$):GOSUB 130
880 PRINT"Fuel moment:                         ",CVI(J$):GOSUB 130
890 PRINT"Forward CG limit:                    ",CVI(K$)
900 PRINT"Aft CG limit:                        ",CVI(L$):GOSUB 130
910 GOSUB 150:GOSUB 140:GOTO 240
920 CLOSE #1:OPEN "R",#1,"WEIGHT.DTA",47
930 FIELD #1,20 AS T$,6 AS U$,4 AS V$,4 AS W$,4 AS X$,4 AS Y$,4 AS Z$
940 GET #1,1
950 PRINT TAB(20)T$:PRINT TAB(20)U$:GOSUB 130
960 PRINT"Key in data in pounds.  For empty type >0<":GOSUB 130
970 INPUT "Pilot weight?                ",PW
980 INPUT "Copilot weight?              ",CW
990 INPUT "Number of passenger seats used?      ",PS
1000 IF PS=0 THEN 1120
1010 INPUT "Row 2, left seat passenger weight?   ",L2
1020 INPUT "Row 2, right seat passenger weight?  ",R2
1030 IF PS<3 THEN 1120
1040 INPUT "Row 3, left seat passenger weight?   ",L3
1050 INPUT "Row 3, right seat passenger weight?  ",R3
1060 IF PS<5 THEN 1120
1070 INPUT "Row 4, left seat passenger weight?   ",L4
1080 INPUT "Row 4, right seat passenger weight?  ",R4
1090 IF PS<7 THEN 1120
1100 INPUT "Row 5, left seat passenger weight?   ",L5
1110 INPUT "Row 5, right seat passenger weight?  ",R5:GOSUB 130
1120 INPUT "Nose compartment baggage weight?     ",NB
1130 INPUT "Aft cabin area baggage weight?       ",AB
1140 INPUT "Rear bay baggage weight?             ",RB
1150 GOSUB 170:IF CORRECT$<>"N" THEN 1160 ELSE GOSUB 140:GOTO 960
1160 TOTPASS=PW+CW+L2+R2+L3+R3+L4+R4+L5+R5
1170 TOTBAGG=NB+AB+RB
1180 PAYLOAD=TOTPASS+TOTBAGG:GOSUB 140
1190 PRINT"Total seat weight:          ";TOTPASS;" pounds.":PRINT
1200 PRINT"Total baggage weight:       ";TOTBAGG;" pounds.":PRINT
1210 PRINT"Payload:                    ";PAYLOAD;" pounds.":GOSUB 130
1220 ZEROFW=PAYLOAD+CVI(V$):IF ZEROFW>CVI(W$) THEN 1370
```

continued

```
1230 PRINT:PRINT"Zero fuel weight:           ";ZEROFW;" pounds.  O.K.":GOSUB 150:GOSUB 140
1240 INPUT "Fuel on board (pounds)?          ",FOB
1250 GOSUB 180:IF CORRECT$<>"N" THEN 1260 ELSE GOSUB 140:GOTO 1240
1260 GROSS=ZEROFW+FOB:IF GROSS>CVI(X$) THEN 1380
1270 PRINT:PRINT"Ramp weight:                ";GROSS;" pounds.  O.K.":GOSUB 150:GOSUB 140
1280 INPUT "Fuel used for taxi and holding?     ",FTH
1290 GOSUB 180:IF CORRECT$<>"N" THEN 1300 ELSE GOSUB 140:GOTO 1280
1300 TAKE=GROSS-FTH:IF TAKE>CVI(Y$) THEN 1390
1310 PRINT:PRINT"Takeoff weight:             ";TAKE;" pounds.  O.K.":PRINT:PRINT:GOSUB 1420
1320 GOSUB 140:INPUT "Estimated fuel burn during trip?     ",EFB
1330 GOSUB 180:IF CORRECT$<>"N" THEN 1340 ELSE GOSUB 140:GOTO 1320
1340 LAND=TAKE-EFB:IF LAND>CVI(Z$) THEN 1400
1350 PRINT:PRINT"Estimated landing weight: ";LAND;" pounds.  O.K."
1360 GOSUB 130:GOSUB 140:GOTO 240
1370 PRINT:PRINT:PRINT ZEROFW;" is above the max zero fuel weight of ";CVI(W$);" pounds."
1380 GOTO 1420
1390 PRINT:PRINT:PRINT GROSS;" is above the max ramp weight of ";CVI(X$);" pounds.":GOTO
1420
1400 PRINT:PRINT:PRINT TAKE;" is above the max takeoff weight of ";CVI(Y$);" pounds.":GOTO
1420
1410 PRINT:PRINT:PRINT LAND;" is above the max landing weight of ";CVI(Z$);" pounds.":GOTO
1420
1420 CLOSE #1:GOSUB 150:GOSUB 140:GOTO 240
1430 OPEN "R",#2,"MOMENT.DTA",46
1440 FIELD #2,6 AS A$,3 AS B$,3 AS C$,3 AS D$,3 AS E$,3 AS F$,3 AS G$,3 AS H$,3 AS I$,5 AS
J$,5 AS K$,5 AS L$
1450 GET #2,1
1460 ROW1=PW+CW:MOM1=ROW1*CVI(B$)
1470 ROW2=L2+R2:MOM2=ROW2*CVI(C$)
1480 ROW3=L3+R3:MOM3=ROW3*CVI(D$)
1490 ROW4=L4+R4:MOM4=ROW4*CVI(E$)
1500 ROW5=L5+R5:MOM5=ROW5*CVI(F$)
1510 MOMN=NB*CVI(G$)
1520 MOMA=AC*CVI(H$)
1530 MOMB=RB*CVI(I$)
1540 MOMF=FOB*CVI(J$)
1550 MOMP=CVI(A$)*1000
1560 TOTMOM=MOM1+MOM2+MOM3+MOM4+MOM5+MOMN+MOMA+MOMB+MOMF+MOMP
1570 CG=TOTMOM/TAKE
1580 CLOSE #2
1590 IF CG<CVI(K$) THEN 1620
1600 IF CG>CVI(L$) THEN 1630
1610 PRINT:PRINT:PRINT"The CG is ";CG;" and is within acceptable limits.":GOSUB 150:RETURN
1620 PRINT:PRINT:PRINT"The CG is ";CG;" and is beyond the forward limit.":PRINT:GOTO 1640
1630 PRINT:PRINT:PRINT"The CG is ";CG;" and is beyond the aft limit.":PRINT
1640 PRINT"Weight must be redistributed!":GOSUB 150:GOSUB 140:GOTO 960
```

Figure 31.1 The weight and balance program that can be used with any aircraft with up to ten seats.

CHAPTER THIRTY-TWO

A Program That Computes Take-Off Distances

The safe takeoff distance, ground run, obstacle-clearance distance and accelerate/stop distance or, in the case of jet aircraft, balanced field length, are determined by temperature, airport elevation, wind direction and velocity, and runway surface conditions. Aircraft operating manuals provide these figures for a variety of conditions, but instead of trying to find one's way through that bunch of data, usually provided in very small print, it's simpler to use this program.

The user must enter specific aircraft data before the program can be used for the first time. These data are stored in a data file and used for all subsequent runs. Thus, prior to the first use of the program, we have to select the second choice from the menu:

```
Do you want to:
---------------------------------------------------------------------
1    Use previously entered aircraft data
2    Enter new aircraft data
---------------------------------------------------------------------
3    Review previously entered aircraft data
---------------------------------------------------------------------
4    Exit program
---------------------------------------------------------------------
Which?
```

Once that choice has been selected, the program asks:

```
Are you sure? Previous entries will be erased! (Y/N)
```

to avoid an accidental loss of previously entered aircraft data. Assuming that we've answered in the affirmative by typing Y, the program continues with a series of prompts:

```
Aircraft identification?
N-number?
Max gross weight?
Ground run, sea level, ISA?
Obstacle distance, sea level, ISA?
Accelerate/stop, sea level, ISA?
```

The figures for the ground run, obstacle distance and accelerate/stop distance under standard temperature conditions (ISA) must be taken from the owner's manual. Once they have been keyed in, the computer stores them in a separate file and uses them to perform the necessary computations.

When you're ready to go on your next flight, activate the program and select the first choice from the above menu. This results in a different series of prompts:

```
Take-off weight?
Airport elevation?
Temperature, degrees C.?
Runway number?
Wind direction?
Wind velocity?
------------------------------------------------------------------------
Runway condition:
------------------------------------------------------------------------
Hard or soft (H or S)?
Smooth or rough (S or R)?
Wet or dry (W or D)?
```

After a very short pause, the program produces the desired results:

```
The density altitude:                                      4328 feet.
------------------------------------------------------------------------
The approximate ground run:                                1875 feet.
The approximate 50-foot obstacle distance:                 2568 feet.
The approximate accelerate/stop distance:                  3106 feet.
------------------------------------------------------------------------
```

In the event that one or several of those distances exceed 5,000 feet, the program adds a warning:

Caution. Possibility of brake fade!

The distances shown are not necessarily correct to the last foot, but they fall within a plus/minus ten percent range. If the situation is critical because of runway length limitations or for any other reason, consult the owner's manual for more precise data.

The third choice from the main menu lets you examine previously entered aircraft data to make sure they are still valid. This selection results in the following display (here using arbitrary figures):

Data for Cessna 441 Conquest
N3371G

Max gross weight:	8925
Ground run, sea level, ISA:	1988
Obstacle distance, sea level, ISA:	2875
Accelerate/stop distance, sea level, ISA:	3780

If you have changed aircraft or if the previously recorded figures must be changed for some other reason, select the second choice from the main menu and enter a new set of figures; these will replace the old ones.

```
100 REM TAKEOFF DISTANCE DATA
110 REM BASIC-80
120 GOTO 180
130 PRINT STRING$(80,45):RETURN
140 HOME:VTAB(10):RETURN
150 PRINT:INPUT "Press >RETURN< (Q to quit)   ",R$:IF R$="Q" THEN 160 ELSE RETURN
160 GOSUB 140:GOSUB 130:PRINT TAB(38)"End.":GOSUB 130:END
170 PRINT:INPUT "Entries correct?  (Y/N)   ",CORRECT$:RETURN
180 GOSUB 140:GOSUB 130
190 PRINT TAB(10)"This program calculates the takeoff distances:":GOSUB 130
200 PRINT TAB(20)"Ground roll":PRINT TAB(20)"50-foot obstacle clearance"
210 PRINT TAB(20)"Accelerate stop distance":GOSUB 130
220 PRINT TAB(10)"The results are accurate within +/-10%.  For precise distance":PRINT
230 PRINT TAB(10)"figures, consult your owners manual.":GOSUB 130
240 PRINT TAB(10)"Before the program can be used for the first time,":PRINT
250 PRINT TAB(10)"you must key in certain data from the owners manual."
260 GOSUB 130:GOSUB 150:GOSUB 140
270 PRINT TAB(20)"Do you want to:":GOSUB 130
280 PRINT 1,"Use previously entered aircraft data"
290 PRINT 2,"Enter new aircraft data.":GOSUB 130
300 PRINT 3,"Review previously entered aircraft data":GOSUB 130
310 PRINT 4,"Exit program":GOSUB 130
320 INPUT "Which?    ",WHICH:GOSUB 140
330 ON WHICH GOTO 600,340,490,160
```

continued

```
340 PRINT"Are you sure?  Previous entries will be erased!";
350 INPUT "  (Y/N)     ",YN$
360 IF YN$<>"Y" THEN GOSUB 140:GOTO 270 ELSE GOSUB 140:GOTO 370
370 OPEN "O",#1,"PLANE.DTA"
380 INPUT "Aircraft identification?              ",AIRCR$
390 PRINT #1,AIRCR$
400 INPUT "N-number?                        ",NUM$
410 PRINT #1,NUM$
420 INPUT "Max gross weight?                 ",GW
430 INPUT "Ground run, sea level, ISA?       ",RG
440 INPUT "Obstacle distance, sea level, ISA?  ",OC
450 INPUT "Accelerate/stop, sea level, ISA?  ",AD
460 GOSUB 170:IF CORRECT$<>"N" THEN 470 ELSE GOSUB 140:GOTO 380
470 PRINT #1,GW;RG;OC;AD
480 CLOSE #1:GOSUB 140:GOTO 270
490 OPEN "I",#1,"PLANE.DTA"
500 INPUT #1,AIRCR$
510 PRINT TAB(20)"Data for ";AIRCR$
520 INPUT #1,NUM$
530 PRINT TAB(20)NUM$:GOSUB 130
540 INPUT #1,GW,RG,OC,AD
550 PRINT"Max gross weight=                       ";GW:PRINT
560 PRINT"Ground run, sea level, ISA=             ";RG:PRINT
570 PRINT"Obstacle distance, sea level, ISA=      ";OC:PRINT
580 PRINT"Accelerate/stop distance, sea level, ISA= ";AD:GOSUB 130
590 CLOSE #1:GOSUB 150:GOTO 270
600 OPEN "I",#1,"PLANE.DTA"
610 INPUT #1,AIRCR$:PRINT TAB(20)AIRCR$
620 INPUT #1,NUM$:PRINT TAB(20)NUM$:GOSUB 130
630 INPUT "Takeoff weight?                   ",TW
640 INPUT "Airport elevation?                ",AE
650 INPUT "Temperature, degrees C.?          ",C
660 Z=288.15:DA=(145426!*(1-(((Z-AE*.001981)/Z)^5.2563/((273.15+C)/Z))^.235))
670 INPUT "Runway number?                    ",RN
680 INPUT "Wind direction?                   ",WD
690 INPUT "Wind velocity?                    ",WV:GOSUB 130
700 PRINT TAB(10)"Runway condition:":GOSUB 130
710 INPUT "Hard or soft     (H or S)?        ",HS$
720 INPUT "Smooth or rough  (S or R)?        ",SR$
730 INPUT "Wet or dry       (W or D)?        ",DW$:GOSUB 130:GOSUB 140
740 GOSUB 170:IF CORRECT$<>"N" THEN 750 ELSE GOSUB 140:GOTO 630
750 INPUT #1,GW,RG,OC,AD
760 WTPER=(GW-TW)/GW*100
770 FOR Y=0 TO 100 STEP 10
780 IF WTPER<=Y THEN 790 ELSE NEXT Y
790 FACTOR3=Y*(WTPER/100)
800 ELEV=.0825
810 TEMP=.12
820 WCOMP=-1*WV*COS((WD-(RN*10))/57.2958)
830 IF HS$="H" THEN RC1=0 ELSE RC1=3
840 IF SR$="S" THEN RC2=0 ELSE RC2=3
850 IF DW$="D" THEN RC3=0 ELSE RC3=3:RCOND=(RC1+RC2+RC3)/100
860 FOR X=1 TO 10000 STEP 1000
870 IF DA<=X THEN 880 ELSE NEXT X
880 FACTOR1=X*ELEV
890 FOR XX=-20 TO 100 STEP 10
900 IF C<=XX THEN 910 ELSE NEXT XX
910 FACTOR2=XX*TEMP
920 RG=RG+WCOMP-FACTOR3+(FACTOR2*.9)+(FACTOR1*.85)+(RG*RCOND):RG=INT(RG+.5)
930 PRINT"The density altitude:                    ";INT(DA+.5);" feet.":GOSUB 130
940 PRINT"The approximate ground run:            ";RG;" feet.":PRINT
950 OC=OC+WCOMP-FACTOR3+(OC*RCOND)+FACTOR1+FACTOR2:OC=INT(OC+.5)
960 PRINT"The approximate 50-foot obstacle distance:   ";OC;" feet.":PRINT
970 AD=AD+WCOMP-FACTOR3+(AD*RCOND)+(FACTOR1*1.65)+(FACTOR2*1.45):AD=INT(AD+.5)
```

```
980 PRINT"The approximate accelerate/stop distance:      ";AD;" feet.":GOSUB 130
990 IF AD>5000 THEN 1000 ELSE 1010
1000 PRINT TAB(10)"Caution. Possibility of brake fade!":GOSUB 130
1010 GOSUB 150:CLOSE #1:GOSUB 140:GOTO 270
```

Figure 32.1 The program that determines takeoff distances based on weather, elevation and runway
condition data.

Lines 100–330 display a description of the program (Figure 32.1), followed by the main menu and a command that sends the computer to one of four line numbers, depending on the selection made by the user.

Lines 340–480 are used to permit entry of the aircraft data, stored in a file called PLANE.DTA. This is a sequential file, meaning that previously stored data are erased when new ones are keyed in.

Lines 490–590 are used to take a look at previously stored data in order to make sure they are still valid.

Lines 600–1010 are used to permit entry of current conditions. They then perform the necessary calculations and display the results.

If the program is to be used for several different aircraft, it should be copied onto separate disks, one for each of the aircraft. Do not make several copies of the program on one disk. Eachcopy would use the same sequential data file unless the file name is changed in each copy of the program.

CHAPTER THIRTY-THREE

Determining the Best Altitude for Cruise

You will find this to be a most useful program. It can be used with any aircraft, single- or twin-engine piston airplanes, turboprops, jets and, yes, even helicoptors. It determines the most advantageous cruising altitude under prevailing upper-wind conditions in terms of economy or time en route.

It determines the amount of fuel that will be burned at any given altitude, and based on time of departure, displays the time of arrival with the time en route. It also displays the average ground speed, including time to climb and descend.

To activate the program type the file name followed by /F:5 (FIL-ENAME.SUB/F:5). The computer must be told that the program produces five separate data files. The program then presents you with the menu, offering a number of choices:

This program determines the best cruising altitude in terms of either economy or time en route, the ground speed and ETA based on expected upper winds.

Menu:

1, Use performance data previously entered.
2, Enter aircraft identification.
3, Enter new performance and/or altitude data.

4, Review previously entered performance or altitude data.

5, Exit program.

Which?

Before the program can be used for the first time, you must enter a number of data (choice number 3 from the menu). These are stored in the group of separate data files and remain there to be used each time the program is run. If you ever want to change those data simply enter the new parameters, automatically erasing the old. This data entry is performed in response to a series of prompts:

```
Enter aircraft designation
N-number?
Percent of power?
```

After these questions have been answered, you'll be presented with a menu of choices about the data to be entered or changed:

```
Enter new data for:
---------------------------------------------------------------------------------
1, Altitudes.
2, Climb and descent rates, TAS and fuel flow.
3, TAS at altitudes.
4, Fuel flow at altitudes.
---------------------------------------------------------------------------------
5, Return to main menu.
---------------------------------------------------------------------------------
Which?
```

The first time the program is run you must enter data for each category. Later, if only certain data are to be changed, just pick the category in question. The prompts for each of the categories are:

```
Enter up to nine cruising altitudes, numbered from 1 to 9.
If less than nine altitudes are to be entered.
type number of altitudes.
```

To use all nine altitudes, simply press the RETURN key; to use less than nine, type the number.

```
Altitude number? (1 to 9)
Altitude in feet?
```

Here type the altitude number in reply to the first prompt and the altitude in feet (3000, 6000 etc.) in answer to the second prompt. This will be repeated for up to nine altitudes.

The next group deals with climb and descent data:

```
Rate of climb? (fpm)
Rate of descent? (fpm)
Climb TAS? (mph or knots)
Descent TAS? (mph or knots)
Climb fuel flow? (pph or gph)
Descent fuel flow? (pph or gph)
```

Finally, there are the true airspeed (TAS) and fuel-flow data at each altitude:

```
Number of altitudes?
TAS at altitude #1?
TAS at altitude #2?
(etc. for each altitude.)
-----------------------------------------------------------------------
Fuel flow at 3000 feet?
Fuel flow at 6000 feet?
(etc. for each altitude.)
```

These data are stored in the five separate data files and used to determine the appropriate time and fuel data for each flight.

If you want to take a look at the data you entered, you may do so by selecting one or several choices from this menu:

```
Review data for:
-----------------------------------------------------------------------
1, Altitudes
2, Climb and descent rates, TAS and fuel flow.
3, TAS at altitudes.
4, Fuel flow at altitudes.
5, Aircraft identification and percent of power.
-----------------------------------------------------------------------
6, Return to main menu.
-----------------------------------------------------------------------
Which?
```

When the program is used to determine the best cruising altitude for a flight, you're asked to answer the following series of questions:

```
Distance to destination?
Magnetic course?
Magnetic variation (E- / W+)
Departure airport elevation?
Destination airport elevation?
ETD? (hrs.min)
Time difference? (E+ / W-)
```

```
Next enter forecast winds. For light and variable, type >1<
Number of altitudes?
```

```
Altitude #1 at 3000 feet:
Wind direction?
Wind velocity?
```

The altitude question will be repeated for each previously entered altitude. The program then goes to work and presents you with the following information for each altitude (the numbers used here are arbitrary):

```
Cruising altitude: 9000 feet.
```

```
Time en route: 3 hours and 47 minutes.
Fuel used: 45.6 gallons or pounds.
```

```
At a true airspeed of 152 your ground speed will be 167
```

```
Estimated time of arrival is 12:42
```

This will be repeated for each altitude, giving you a chance to compare the data to decide which will be the most economical or fastest.

The program (Figure 33.1) consists of 223 lines divided into five groups. Each group performs certain tasks. The first group (lines 100–290) displays the purpose of the program and the main menu. It sends the computer to one of five line numbers, depending on the selection made by the user.

```
100 REM ALTITUDE SELECTION PROGRAM.  FILENAME.SUB/F:5
110 REM BASIC-80.
120 GOTO 180
130 PRINT STRING$(80,45):RETURN
140 HOME:VTAB(10):RETURN
150 PRINT:INPUT "Press >RETURN<  (Q to quit)  ",R$:IF R$="Q" THEN 160 ELSE RETURN
160 GOSUB 140:GOSUB 130:PRINT TAB(38)"End.":GOSUB 130:END
170 PRINT:INPUT "Entries correct?  (Y/N)  ",CORRECT$:RETURN
180 ZZ=0:GOSUB 140:GOSUB 130
190 PRINT TAB(7)"This program determines the best cruising altitude in terms of either":PRINT
200 PRINT TAB(7)"economy or time en route, the ground speed and ETA based on expected":PRINT
210 PRINT TAB(7)"upper winds.":GOSUB 130:GOSUB 150:GOSUB 140
220 PRINT TAB(10)"Menu:":GOSUB 130
230 PRINT 1,"Use performance data previously entered."
240 PRINT 2,"Enter aircraft identification."
250 PRINT 3,"Enter new performance and/or altitude data.":GOSUB 130
260 PRINT 4,"Review perviously entered performance and/or altitude data.":GOSUB 130
270 PRINT 5,"Exit program.":GOSUB 130
280 INPUT "Which? ",WHICH:GOSUB 140
290 ON WHICH GOTO 1440,300,420,1020,160
300 INPUT "Are you sure.  Previous entries will be erased!  (Y/N)  ",ERSE$
310 IF ERSE$<>"N" THEN GOSUB 140:GOTO 320 ELSE GOSUB 140:GOTO 220
320 OPEN "R",#5,"AIRCRAFT.DTA",30
330 FIELD #5,20 AS A$,6 AS B$,3 AS C$
340 INPUT "Enter aircraft designation            ",AIRCRAFT$
350 INPUT "N-number?                             ",NUMBER$
360 INPUT "Percent of power?                     ",POWER%
370 GOSUB 170:IF CORRECT$<>"N" THEN 380 ELSE GOSUB 140:GOTO 340
380 LSET A$=AIRCRAFT$
390 LSET B$=NUMBER$
400 LSET C$=MKI$(POWER%)
410 PUT #5,1:CLOSE #5:GOSUB 140:GOTO 220
420 INPUT "Are you sure.  Previous entries will be erased!  (Y/N)  ",ERSE$
430 IF ERSE$<>"N" THEN GOSUB 140:GOTO 440 ELSE GOSUB 140:GOTO 220
440 PRINT TAB(10)"Enter new data for:":GOSUB 130
450 PRINT 1,"Altitudes."
460 PRINT 2,"Climb and descent rates, TAS and fuel flow."
470 PRINT 3,"TAS at altitudes."
480 PRINT 4,"Fuel flow at altitudes.":GOSUB 130
490 PRINT 5,"Return to main menu.":GOSUB 130
500 INPUT "Which?  ",WHICH:GOSUB 140
510 ON WHICH GOTO 520,640,800,900,220
520 OPEN "R",#1,"ALTITUDE.DTA",6
530 FIELD #1,5 AS AA$
540 PRINT"Enter up to nine cruising altitudes, numbered from 1 to 9.":PRINT
550 PRINT"If less than nine altitudes are to be entered,":PRINT
560 INPUT "type number of altitudes              ",NUMALT%:GOSUB 130
570 FOR ALTNUM%=1 TO NUMALT%:PRINT"Altitude number              ";ALTNUM%
580 INPUT "Altitude in feet?                     ",ALTFT(ALTNUM%):PRINT:PRINT
590 LSET AA$=MKI$(ALTFT(ALTNUM%))
600 PUT #1,ALTNUM%
610 NEXT ALTNUM%
620 GOSUB 170:IF CORRECT$<>"N" THEN 630 ELSE GOSUB 140:GOTO 540
630 CLOSE #1:GOSUB 140:GOTO 420
640 OPEN "R",#2,"CLIMBDES.DTA",25
650 FIELD #2,4 AS C$,4 AS D$,4 AS T$,4 AS S$,4 AS F$,4 AS G$
660 INPUT "Rate of climb? (fpm)                  ",ROC
670 INPUT "Rate of descent? (fpm)                ",DES
680 INPUT "Climb TAS? (mph or knots)             ",CTAS
690 INPUT "Descent TAS? (mph or knots)           ",DTAS
700 INPUT "Climb fuel flow? (pph or gph)         ",CFF
710 INPUT "Descent fuel flow? (pph or gph)       ",DFF
720 GOSUB 170:IF CORRECT$<>"N" THEN 730 ELSE GOSUB 140:GOTO 660
```

```
730 LSET C$=MKI$(ROC)
740 LSET D$=MKI$(DES)
750 LSET T$=MKI$(CTAS)
760 LSET S$=MKI$(DTAS)
770 LSET F$=MKI$(CFF)
780 LSET G$=MKI$(DFF)
790 PUT #2,1:CLOSE #2:GOSUB 140:GOTO 420
800 OPEN "R",#3,"TAS.DTA",28
810 FIELD #3,3 AS A$,3 AS B$,3 AS C$,3 AS D$,3 AS E$,3 AS F$,3 AS G$,3 AS H$,3 AS I$
820 INPUT "Number of altitudes?                    ",NUMALT%
830 FOR X=1 TO NUMALT%
840 PRINT"TAS at altitude #";X:INPUT ALT(X)
850 NEXT X
860 LSET A$=MKI$(ALT(1)):LSET B$=MKI$(ALT(2)):LSET C$=MKI$(ALT(3))
870 LSET D$=MKI$(ALT(4)):LSET E$=MKI$(ALT(5)):LSET F$=MKI$(ALT(6))
880 LSET G$=MKI$(ALT(7)):LSET H$=MKI$(ALT(8)):LSET I$=MKI$(ALT(9))
890 PUT #3,1:CLOSE #3:GOSUB 140:GOTO 420
900 OPEN "R",#4,"FUEL.DTA",28
910 FIELD #4,3 AS A$,3 AS B$,3 AS C$,3 AS D$,3 AS E$,3 AS F$,3 AS G$,3 AS H$,3 AS I$
920 INPUT "Number of altitudes?                    ",NUMALT%
930 FOR Y=1 TO NUMALT%
940 PRINT "Fuel flow at ";ALTFT(Y);" feet  ";
950 INPUT FF(Y)
960 NEXT Y
970 GOSUB 170:IF CORRECT$<>"N" THEN 980 ELSE GOSUB 140:GOTO 920
980 LSET A$=MKI$(FF(1)):LSET B$=MKI$(FF(2)):LSET C$=MKI$(FF(3))
990 LSET D$=MKI$(FF(4)):LSET E$=MKI$(FF(5)):LSET F$=MKI$(FF(6))
1000 LSET G$=MKI$(FF(7)):LSET H$=MKI$(FF(8)):LSET I$=MKI$(FF(9))
1010 PUT #4,1:CLOSE #4:GOSUB 140:GOTO 220
1020 PRINT TAB(10)"Review data for:":GOSUB 130
1030 PRINT 1,"Altitudes."
1040 PRINT 2,"Climb and descent rates, TAS and fuel flow."
1050 PRINT 3,"TAS at altitudes."
1060 PRINT 4,"Fuel flow at altitudes.":GOSUB 130
1070 PRINT 5,"Aircraft identification and percent of power.":GOSUB 130
1080 PRINT 6,"Return to main menu.":GOSUB 130
1090 INPUT "Which?  ",WHICH:GOSUB 140:ZZ=5
1100 ON WHICH GOTO 1110,1190,1290,1380,1440,220
1110 OPEN "R",#1,"ALTITUDE.DTA",6
1120 FIELD #1,5 AS A$
1130 FOR X=1 TO 9
1140 GET #1,X
1150 ALT(X)=CVI(A$)
1160 PRINT TAB(30)ALT(X)
1170 NEXT X
1180 CLOSE #1:GOSUB 150:GOSUB 140:GOTO 1020
1190 OPEN "R",#2,"CLIMBDES.DTA",25
1200 FIELD #2,4 AS C$,4 AS D$,4 AS T$,4 AS S$,4 AS F$,4 AS G$
1210 GET #2,1
1220 PRINT TAB(20)"Rate of climb:      ";CVI(C$)
1230 PRINT TAB(20)"Rate of descent:    ";CVI(D$)
1240 PRINT TAB(20)"Climb TAS:          ";CVI(T$)
1250 PRINT TAB(20)"Descent TAS:        ";CVI(S$)
1260 PRINT TAB(20)"Climb fuel flow:    ";CVI(F$)
1270 PRINT TAB(20)"Descent fuel flow:  ",CVI(G$)
1280 CLOSE #2:GOSUB 130:GOSUB 150:GOSUB 140:GOTO 1020
1290 TT$="TAS at altitude #"
1300 OPEN "R",#3,"TAS.DTA",28
1310 FIELD #3,3 AS A$,3 AS B$,3 AS C$,3 AS D$,3 AS E$,3 AS F$,3 AS G$,3 AS H$,3 AS I$
1320 GET #3,1
1330 PRINT TAB(20)TT$;"1 = ";CVI(A$):PRINT TAB(20)TT$;"2 = ";CVI(B$)
1340 PRINT TAB(20)TT$;"3 = ";CVI(C$)
1350 PRINT TAB(20)TT$;"4 = ";CVI(D$):PRINT TAB(20)TT$;"5 = ";CVI(E$)
```

continued

```
1360 PRINT TAB(20)TT$;"6 = ";CVI(F$)
1370 PRINT TAB(20)TT$;"7 = ";CVI(G$):PRINT TAB(20)TT$;"8 = ";CVI(H$)
1380 PRINT TAB(20)TT$;"9 = ";CVI(I$)
1390 IF ZZ=5 THEN 1460 ELSE 1400
1400 CLOSE #3:GOSUB 130:GOSUB 150:GOSUB 140:GOTO 1020
1410 FF$="Fuel flow at altitude #"
1420 OPEN "R",#4,"FUEL.DTA",28
1430 FIELD #4,3 AS A$,3 AS B$,3 AS C$,3 AS D$,3 AS E$,3 AS F$,3 AS G$,3 AS H$,3 AS I$
1440 GET #4,1
1450 TT$=FF$:ZZ=5:GOTO 1330
1460 CLOSE #4:GOSUB 130:GOSUB 150:GOSUB 140:GOTO 1020
1470 OPEN "R",#5,"AIRCRAFT.DTA",30
1480 FIELD #5,20 AS A$,6 AS B$,3 AS C$
1490 GET #5,1
1500 PRINT TAB(20)A$,B$
1510 PRINT TAB(20)"at ";CVI(C$);"% of power":GOSUB 130
1520 CLOSE #5:IF WHICH=1 THEN 1540
1530 IF ZZ=5 THEN GOSUB 150:GOTO 1020 ELSE 1540
1540 INPUT "Distance to destination          ",DIST
1550 INPUT "Magnetic course?                 ",MAG
1560 INPUT "Magnetic variation? (E- / W+)    ",VAR
1570 INPUT "Departure airport elevation?     ",DEP
1580 INPUT "Destination airport elevation?   ",DES
1590 INPUT "ETD?  (hrs.min)                  ",ETD
1600 INPUT "Time difference?     (E+ / W-)   ",DIF:GOSUB 130
1610 GOSUB 170:IF CORRECT$<>"N" THEN 1620 ELSE GOSUB 140:GOTO 1540
1620 GOSUB 140
1630 PRINT "Next enter forecast winds. For light and variable, type >1<":PRINT
1640 INPUT "Number of altitudes?             ",NUMALT%:GOSUB 130
1650 ZZ=3:GOTO 1730
1660 FOR Z=1 TO NUMALT%
1670 ALT=ALT(Z):GOSUB 130
1680 PRINT"Altitude #";Z;" at ";ALT;" feet":PRINT
1690 INPUT "Wind direction?                  ",WD(Z)
1700 INPUT "Wind velocity?                   ",WV(Z)
1710 ZZ=0:NEXT Z
1720 GOSUB 170:IF CORRECT$<>"N" THEN 1730 ELSE GOSUB 140:GOTO 1660
1730 OPEN "R",#1,"ALTITUDE.DTA",6
1740 FIELD #1,5 AS AA$
1750 FOR X=1 TO 9
1760 GET #1,X
1770 ALT(X)=CVI(AA$)
1780 NEXT X:CLOSE #1:IF ZZ=3 THEN 1660 ELSE 1790
1790 OPEN "R",#2,"CLIMBDES.DTA",25
1800 FIELD #2,4 AS C$,4 AS D$,4 AS T$,4 AS S$,4 AS F$,4 AS G$
1810 GET #2,1
1820 CLOSE #2
1830 CLIMB=CVI(C$):DESC=CVI(C$):CTAS=CVI(T$):DTAS=CVI(S$):CFUL=CVI(F$):DFUL=CVI(G$)
1840 OPEN "R",#3,"TAS.DTA",28
1850 FIELD #3,3 AS A$,3 AS B$,3 AS C$,3 AS D$,3 AS E$,3 AS F$,3 AS G$,3 AS H$,3 AS I$
1860 GET #3,1
1870 CLOSE #3
1880 TAS(1)=CVI(A$):TAS(2)=CVI(B$):TAS(3)=CVI(C$):TAS(4)=CVI(D$):TAS(5)=CVI(E$)
1890 TAS(6)=CVI(F$):TAS(7)=CVI(G$):TAS(8)=CVI(H$):TAS(9)=CVI(I$)
1900 OPEN "R",#4,"FUEL.DTA",28
1910 FIELD #4,3 AS A$,3 AS B$,3 AS C$,3 AS D$,3 AS E$,3 AS F$,3 AS G$,3 AS H$,3 AS I$
1920 GET #4,1
1930 CLOSE #4
1940 FUL(1)=CVI(A$):FUL(2)=CVI(B$):FUL(3)=CVI(C$):FUL(4)=CVI(D$):FUL(5)=CVI(E$)
1950 FUL(6)=CVI(F$):FUL(7)=CVI(G$):FUL(8)=CVI(H$):FUL(9)=CVI(I$)
1960 FOR PASS=1 TO NUMALT%
1970 ALT=ALT(PASS)
1980 GOSUB 140:PRINT"Cruising altitude: ";ALT;" feet.":GOSUB 130
1990 CLIMB=ALT-DEP:CLIMB=CLIMB/CTAS:CLIMB=CLIMB/.6/100
2000 CDIST=CTAS*CLIMB
```

```
2010 DESC=ALT-DES:DESC=DESC/DTAS:DESC=DESC/.6/100
2020 DDIST=DTASXDESC
2030 IF DDIST+CDIST>=DIST THEN 2220 ELSE 2040
2040 TAS=TAS(PASS):WD=WD(PASS):WV=WV(PASS):FUL=FUL(PASS)
2050 WC=-1XWVXCOS((WD-MC-MV)/57.2958):TAS=TAS-WC
2060 TOTDIST=DIST-CDIST-DDIST:TIME=TOTDIST/TAS+CLIMB+DESC
2070 TIME1=INT(TIME):TIME2=TIME-TIME1:TIME3=TIME2X.6X100:TIME3=INT(TIME3)
2080 PRINT"Time en route: ";TIME1;" hours and ";TIME3;" minutes."
2090 FUEL=(CLIMBXCFUL)+(DESCXDFUL)+(TIMEXFUL(PASS)):FUEL=INT(FUELX10+.5)/10
2100 PRINT"Fuel used: ";FUEL;" gallons or pounds.":GOSUB 130
2110 ETD1=INT(ETD):ETD2=ETD-ETD1:ETD3=ETD2X100
2120 ETA1=ETD+DIF+TIME1:ETA2=ETD3+TIME3
2130 IF ETA2>60 THEN ETA2=ETA2-60 AND ETA1=ETA1+1
2140 IF ETA1>24 THEN ETA1=ETA1-24
2150 PRINT"At a true airspeed of ";TAS(PASS);" your "
2160 TAS=INT(TAS):PRINT"ground speed will be ";TAS
2170 ETA1=INT(ETA1):ETA2=INT(ETA2)
2180 PRINT"Estimated time of arrival is ";ETA1;":";ETA2:GOSUB 130
2190 GOSUB 150
2200 NEXT PASS
2210 GOSUB 140:GOTO 220
2220 GOSUB 140:PRINT TAB(20)ALT;" is too high for ";DIST;" miles."
2230 GOSUB 130:GOSUB 150:GOSUB 140:GOTO 220
```

Figure 33.1 The program can be used with any aircraft to determine the most advantageous cruise altitude for any flight under prevailing or forecast upper wind conditions.

The second group (lines 300–410) is used to key in the aircraft identification, N-number and the percent of power normally used in cruise. If you want to use the program for different power percentages, simply copy it onto another disk (it must be another disk to avoid destroying the data in the data files) and enter the appropriate power percentage. Once the entry is complete, the program creates a file called AIR-CRAFT.DTA to store the data for future use.

The third group (lines 420–1010) is used for entry of the number of altitudes to be included, the altitudes in feet, and a variety of performance data. Once this information has been keyed in, it is stored in several files called ALTITUDE.DTA, TAS.DTA, FUEL.DTA and CLIMBDES.DTA, where it remains unless altered or erased by the user.

The fourth group (lines 1020–1400) accesses the data files and displays the previously stored information to permit the user to determine whether or not these data are still valid.

The fifth group (lines 1410–2230) also accesses the data files, using the stored information along with keyed-in data to perform the calculations that produce the results for a specific flight.

If you would like to have these results sent to the line printer, you'll have to change line 2200 to read:

```
2070 GOSUB 2240:NEXT PASS
```

and then add the following lines:

```
2240 PRINT:INPUT "Printout?  (Y/N)  ",PR$
2250 IF PR$<>"Y" THEN 2340 ELSE 2260
2260 LPRINT:LPRINT "Cruising altitude: ";ALT;" feet."
2270 LPRINT STRING$(50,45)
2280 LPRINT "Time en route: ";TIME1;" hours and ";TIME3;" minutes"
2290 LPRINT "Fuel used:     ";FUEL;" gallons or pounds"
2300 LPRINT "At a true airspeed of ";TAS(PASS);" your"
2310 LPRINT "ground speed will be ";TAS
2320 LPRINT "Estimated time of arrival is ";ETA1;":";ETA2
2330 LPRINT STRING$(50,45)
2340 RETURN
```

The advantage of making this small addition is that it will simplify the task of comparing the time and fuel data for different altitudes, especially if you are examining all nine altitudes (Figure 33.2).

```
Cruising altitude:  3000  feet.
--------------------------------------------------
Time en route:  6  hours and  44  minutes
Fuel used:        90.8  gallons or pounds
At a true airspeed of  130  your
ground speed will be  118
Estimated time of arrival is  13 : 0
--------------------------------------------------

Cruising altitude:  6000  feet.
--------------------------------------------------
Time en route:  6  hours and  28  minutes
Fuel used:        89.9  gallons or pounds
At a true airspeed of  135  your
ground speed will be  120
Estimated time of arrival is  13 : 58
--------------------------------------------------

Cruising altitude:  9000  feet.
--------------------------------------------------
Time en route:  6  hours and  8  minutes
Fuel used:        88.7  gallons or pounds
At a true airspeed of  140  your
ground speed will be  127
Estimated time of arrival is  13 : 38
--------------------------------------------------

Cruising altitude:  12000  feet.
--------------------------------------------------
Time en route:  5  hours and  54  minutes
Fuel used:        95.2  gallons or pounds
At a true airspeed of  145  your
ground speed will be  132
Estimated time of arrival is  12 : 0
--------------------------------------------------

Cruising altitude:  15000  feet.
--------------------------------------------------
Time en route:  5  hours and  16  minutes
Fuel used:        91.5  gallons or pounds
At a true airspeed of  150  your
ground speed will be  164
Estimated time of arrival is  12 : 46
--------------------------------------------------
```

Figure 33.2 A sample print-out produced by the suggested print option added to the altitude selection for a flight from Santa Fe, New Mexico to Los Angeles.

CHAPTER THIRTY-FOUR

Twelve Flight Data Programs

This rather lengthy utility program performs all the frequently-needed flight-related calculations. Its functions are described by the main menu:

```
MENU:
--------------------------------------------------------------------------------
  1  DISTANCE DATA
  2  GROUND SPEED DATA
  3  TIME EN ROUTE DATA
  4  MILES PER GALLON/POUND DATA
  5  FUEL QUANTITY DATA
  6  FUEL FLOW DATA
  7  FLIGHT PATH ANGLE DATA
  8  FPM CLIMB/DESCENT DATA
  9  TIME TO CLIMB/DESCEND DATA
 10  POINT OF DESCENT DATA
 11  GRADIENT DATA
 12  ALTITUDE CHANGE DATA
 13  EXIT PROGRAM
--------------------------------------------------------------------------------
  YOUR CHOICE BY NUMBER
```

Of these programs, the first six offer an additional choice concerning the data to be used in the calculations:

1   F I ND D I STANCE BASED ON GROUND SPEED AND T I ME EN ROUTE
2   F I ND D I STANCE BASED ON FUEL FLOW AND FUEL ON BOARD

1   F I ND GROUND SPEED BASED ON D I STANCE FLOWN & T I ME EN ROUTE
2   F I ND GROUND SPEED BASED ON FUEL DATA

1   T I ME EN ROUTE BASED ON D I STANCE FLOWN AND GROUND SPEED
2   T I ME EN ROUTE BASED ON FUEL DATA

1   M I LES PER GALLON OR PER POUND BASED ON FUEL BURNED AND M I LES FLOWN
2   M I LES PER GALLON OR PER POUND BASED ON GROUND SPEED AND FUEL FLOW

1   F I ND FUEL NEEDED/CONSUMED BASED ON T I ME EN ROUTE AND FUEL FLOW
2   F I ND FUEL NEEDED/CONSUMED BASED ON M I LES, GROUND SPEED, FUEL FLOW

1   F I ND FUEL FLOW BASED ON T I ME EN ROUTE AND FUEL CON-SUMED
2   F I ND FUEL FLOW BASED ON M I LES FLOWN, GROUND SPEED, FUEL CON-SUMED

In each case you're prompted for the appropriate input, after which the programs display the results.

Lines 100–360 of the program (Figure 34.1) present the menu and send the computer to one of 13 line numbers. The rest of the program consists of the input lines, calculations and result displays for the 12 subprograms.

```
100 REM FLIGHT DATA (12 PROGRAMS)
110 REM BASIC-80
120 GOTO 200
130 PRINT STRING$(80,45):RETURN
140 HOME:VTAB(10):RETURN
150 PRINT:INPUT "Press >RETURN< (Q to quit)  ",Y$:IF Y$="Q" THEN 160 ELSE RETURN
160 GOSUB 140:GOSUB 130:PRINT TAB(38)"End.":GOSUB 130:END
170 INPUT "For another calculation press >RETURN< (Q to quit)  ",Y$
180 IF Y$="Q" THEN 160 ELSE GOTO 210
190 PRINT:INPUT "Entry correct?  (Y/N)  ",CORRECT$:RETURN
200 GOSUB 140:GOSUB 130:PRINT TAB(28)"FLIGHT DATA PROGRAMS":GOSUB 130:GOSUB 150
210 GOSUB 140:VTAB(5):PRINT"MENU:":GOSUB 130
220 PRINT 1,"DISTANCE DATA"
230 PRINT 2,"GROUND SPEED DATA"
240 PRINT 3,"TIME EN ROUTE DATA"
250 PRINT 4,"MILES PER GALLON/POUND DATA"
```

```
260 PRINT 5,"FUEL QUANTITY DATA"
270 PRINT 6,"FUEL FLOW DATA"
280 PRINT 7,"FLIGHT PATH ANGLE DATA"
290 PRINT 8,"FPM CLIMB/DESCENT DATA"
300 PRINT 9,"TIME TO CLIMB/DESCEND DATA"
310 PRINT 10,"POINT OF DESCENT DATA"
320 PRINT 11,"GRADIENT DATA"
330 PRINT 12,"ALTITUDE CHANGE DATA"
340 GOSUB 130:PRINT 13,"EXIT PROGRAM":GOSUB 130
350 INPUT "YOUR CHOICE BY NUMBER ",A:GOSUB 140
360 ON A GOTO 370,550,730,890,1050,1230,1410,1480,1550,1610,1680,1740,160
370 GOSUB 140:PRINT 1,"FIND DISTANCE BASED ON GROUND SPEED AND TIME EN ROUTE"
380 PRINT 2,"FIND DISTANCE BASED ON FUEL FLOW AND FUEL ON BOARD":PRINT
390 INPUT "Which? ",WHICH:GOSUB 140
400 ON WHICH GOTO 410,480
410 INPUT "KEY IN GROUND SPEED IN MPH OR KNOTS ",GS
420 INPUT "TIME EN ROUTE IN HOURS.MINUTES        ",HM
430 GOSUB 190:IF CORRECT$<>"N" THEN 440 ELSE GOSUB 140:GOTO 410
440 H=INT(HM):M=HM-H:M=M/.6:HM=H+M
450 D=GS*HM:D=CINT(D)
460 PRINT:PRINT:GOSUB 130:PRINT"DISTANCE: ";D;" SM OR NM":GOSUB 130:PRINT:PRINT
470 GOTO 170
480 INPUT "FUEL ON BOARD (GALS. OR POUNDS) ",FB:PRINT
490 INPUT "FUEL FLOW (GPH OR PPH)        ",FF:PRINT
500 INPUT "GROUND SPEED (MPH OR KNOTS)   ",GS:PRINT
510 GOSUB 190:IF CORRECT$<>"N" THEN 520 ELSE GOSUB 140:GOTO 480
520 D=(FB/FF)*GS:D=CINT(D)
530 GOSUB 130:PRINT"RANGE:   ";D;" SM OR NM":GOSUB 130:PRINT:PRINT
540 GOTO 170
550 PRINT 1,"FIND GROUND SPEED BASED ON DISTANCE FLOWN & TIME EN ROUTE"
560 PRINT 2,"FIND GROUND SPEED BASED ON FUEL DATA":GOSUB 130
570 INPUT "Which? ",WHICH:GOSUB 140
580 ON WHICH GOTO 590,660
590 INPUT "KEY IN MILES FLOWN              ",D
600 INPUT "TIME EN ROUTE IN HOURS.MINUTES   ",HM:PRINT
610 GOSUB 190:IF CORRECT$<>"N" THEN 620 ELSE GOSUB 140:GOTO 590
620 H=INT(HM):M=HM-H:M=M/.6:HM=H+M
630 GS=D/HM:GS=CINT(GS)
640 GOSUB 130:PRINT"GROUND SPEED: ";GS;" MPH OR KNOTS":GOSUB 130
650 PRINT:GOTO 170
660 INPUT "FUEL FLOW (GPH OR PPH)          ",FF:PRINT
670 INPUT "FUEL BURNED (GAL OR LBS)        ",FB:PRINT
680 INPUT "MILES FLOWN (SM OR NM)          ",D:PRINT
690 GOSUB 190:IF CORRECT$<>"N" THEN 700 ELSE GOSUB 140:GOTO 660
700 GS=D/(FB/FF):GS=CINT(GS)
710 GOSUB 130:PRINT"GROUND SPEED: ";GS;" MPH OR KNOTS":GOSUB 130
720 PRINT:GOTO 170
730 PRINT 1,"TIME EN ROUTE BASED ON DISTANCE FLOWN AND GROUND SPEED"
740 PRINT 2,"TIME EN ROUTE BASED ON FUEL DATA":GOSUB 130
750 INPUT "Which?   ",WHICH:GOSUB 140
760 ON WHICH GOTO 770,830
770 INPUT "GROUND SPEED (MPH OR KNOTS)?      ",GS
780 INPUT "MILES FLOWN (SM OR NM)?          ",D:PRINT
790 GOSUB 190:IF CORRECT$<>"N" THEN 800 ELSE GOSUB 140:GOTO 770
800 HM=D/GS:H=INT(HM):M=HM-H:M=M*.6:HM=H+M
810 GOSUB 130:PRINT"TIME EN ROUTE (HRS.MIN) : ";HM:GOSUB 130
820 PRINT:GOTO 170
830 INPUT "FUEL FLOW (GPH OR PPH)          ",FF
840 INPUT "FUEL BURNED (GAL OR LBS)        ",FB:PRINT
850 GOSUB 190:IF CORRECT$<>"N" THEN 860 ELSE GOSUB 140:GOTO 830
860 HM=FB/FF:H=INT(HM):M=HM-H:M=M*.6:HM=H+M
870 GOSUB 130:PRINT"TIME EN ROUTE (HRS.MIN) :   ";HM:GOSUB 130
880 PRINT:GOTO 170
```

continued

```
890 PRINT 1,"FIND MILES PER GALLON OR PER POUND BASED ON FUEL BURNED AND MILES FLOWN"
900 PRINT 2,"FIND MILES PER GALLON OR PER POUND BASED ON GROUND SPEED AND FUEL FLOW"
910 GOSUB 130:INPUT "Which?    ",WHICH:GOSUB 140
920 ON WHICH GOTO 930,990
930 INPUT "MILES FLOWN?                         ",FF
940 INPUT "FUEL BURNED (GAL OR LBS)?            ",FB:PRINT
950 GOSUB 190:IF CORRECT$<>"N" THEN 960 ELSE GOSUB 140:GOTO 930
960 GP=FF/FB:GP=CINT(GP)
970 GOSUB 130:PRINT"MILES PER GALLON/POUND: ";GP:GOSUB 130
980 PRINT:GOTO 170
990 INPUT "FUEL FLOW (GPH OR PPH)?             ",FF:PRINT
1000 INPUT "GROUND SPEED (MPH OR KNOTS?         ",GS:PRINT
1010 GOSUB 190:IF CORRECT$<>"N" THEN 1020 ELSE GOSUB 140:GOTO 990
1020 GP=GS/FF:GP=CINT(GP)
1030 GOSUB 130:PRINT"MILES PER GALLON/POUND: "GP:GOSUB 130
1040 PRINT:GOTO 170
1050 PRINT 1,"FIND FUEL NEEDED/COMSUMED BASED ON TIME EN ROUTE AND FUEL FLOW"
1060 PRINT 2,"FIND FUEL NEEDED/CONSUMED BASED ON MILES, GROUND SPEED, FUEL FLOW"
1070 GOSUB 130:INPUT "Which?    ",WHICH:GOSUB 140
1080 ON WHICH GOTO 1090,1160
1090 INPUT "FUEL FLOW (GPH OR PPH)?             ",FF
1100 INPUT "TIME EN ROUTE IN HRS.MIN?           ",HM:PRINT
1110 GOSUB 190:IF CORRECT$<>"N" THEN 1120 ELSE GOSUB 140:GOTO 1090
1120 H=INT(HM):M=HM-H:M=M/.6:HM=H+M
1130 FB=FF*HM:FB=INT(FB*10+.5)/10
1140 GOSUB 130:PRINT"FUEL NEEDED/CONSUMED: ";FB;" GALLONS OR POUNDS"
1150 GOSUB 130:PRINT:GOTO 170
1160 INPUT "MILES FLOWN (SM OR NM)?             ",D:PRINT
1170 INPUT "FUEL FLOW (GPH OR PPH)?             ",FF:PRINT
1180 INPUT "GROUND SPEED (MPH OR KTS)?          ",GS:PRINT
1190 GOSUB 190:IF CORRECT$<>"N" THEN 1200 ELSE GOSUB 140:GOTO 1160
1200 FB=(D/GS)*FF:FB=INT(FB*10+.5)/10
1210 GOSUB 130:PRINT"FUEL NEEDED/CONSUMED: ";FB;" GALLONS OR POUNDS"
1220 GOSUB 130:PRINT:GOTO 170
1230 PRINT 1,"FIND FUEL FLOW BASED ON TIME EN ROUTE AND FUEL CONSUMED"
1240 PRINT 2,"FIND FUEL FLOW BASED ON MILES FLOWN, GROUND SPEED, FUEL CONSUMED"
1250 GOSUB 130:INPUT "Which?    ",WHICH:GOSUB 140
1260 ON WHICH GOTO 1270,1340
1270 INPUT "TIME EN ROUTE (HRS.MIN)?            ",HM
1280 INPUT "FUEL CONSUMED (GAL OR LBS)?         ",FB
1290 GOSUB 190:IF CORRECT$<>"N" THEN 1300 ELSE GOSUB 140:GOTO 1270
1300 H=INT(HM):M=HM-H:M=M/.6:M=H+M
1310 FF=FB/HM:FF=CINT(FF)
1320 GOSUB 130:PRINT"FUEL FLOW: ";FF;" GPH OR PPH":GOSUB 130
1330 PRINT:GOTO 170
1340 INPUT "MILES FLOWN (SM OR NM)?             ",D:PRINT
1350 INPUT "GROUND SPEED (MPH OR KTS)?          ",GS:PRINT
1360 INPUT "FUEL CONSUMED (GAL OR LBS)?         ",FB:PRINT
1370 GOSUB 190:IF CORRECT$<>"N" THEN 1380 ELSE GOSUB 140:GOTO 1340
1380 FF=FB/(D/GS):FF=CINT(FF)
1390 GOSUB 130:PRINT"FUEL FLOW: ";FF;" GPH OR PPH":GOSUB 130
1400 PRINT:GOTO 170
1410 INPUT "ALTITUDE CHANGE IN FEET?            ",AC:PRINT
1420 INPUT "HORIZONTAL DISTANCE IN NM?          ",HD:PRINT
1430 GOSUB 190:IF CORRECT$<>"N" THEN 1440 ELSE GOSUB 140:GOTO 1410
1440 HD=HD*6076
1450 A=(ATN(HD/AC))*(180/3.14159):AN=CINT(90-A)
1460 GOSUB 130:PRINT"FLIGHT PATH ANGLE: ";AN;" DEGREES":GOSUB 130
1470 PRINT:GOTO 170
1480 INPUT "ALTITUDE CHANGE IN FEET?            ",AC:PRINT
1490 INPUT "HORIZONTAL DISTANCE IN NM?          ",HD:PRINT
1500 INPUT "GROUND SPEED IN KNOTS?              ",GS:PRINT
1510 GOSUB 190:IF CORRECT$<>"N" THEN 1520 ELSE GOSUB 140:GOTO 1480
1520 FP=HD/GS:FP=FP*60:FP=AC/FP:FP=CINT(FP)
```

```
1530 GOSUB 130:PRINT"RATE OF CLIMB/DESCENT: ";FP;" FPM":GOSUB 130
1540 PRINT:GOTO 170
1550 INPUT "ALTITUDE CHANGE IN FEET?          ",AC:PRINT
1560 INPUT "RATE OF CLIMB/DESCENT?            ",RC:PRINT
1570 GOSUB 190:IF CORRECT$<>"N" THEN 1580 ELSE GOSUB 140:GOTO 1550
1580 HM=AC/RC:HM=CINT(HM)
1590 GOSUB 130:PRINT"TIME TO CLIMB/DESCENT: ";HM;" MINUTES"
1600 GOSUB 130:PRINT:GOTO 170
1610 INPUT "ALTITUDE CHANGE IN FEET?          ",AC:PRINT
1620 INPUT "RATE OF CLIMB/DESCENT?            ",RC:PRINT
1630 INPUT "GROUND SPEED IN KNOTS?            ",GS:PRINT
1640 GOSUB 190:IF CORRECT$<>"N" THEN 1650 ELSE GOSUB 140:GOTO 1610
1650 PD=AC/RC/60%GS:PD=CINT(PD)
1660 GOSUB 130:PRINT"POINT OF DESCENT: ";PD;" NM FROM DESTINATION"
1670 GOSUB 130:PRINT:GOTO 170
1680 INPUT "ALTITUDE CHANGE IN FEET PER NM?  ",AC:PRINT
1690 INPUT "GROUND SPEED IN KNOTS?           ",GS:PRINT
1700 GOSUB 190:IF CORRECT$<>"N" THEN 1710 ELSE GOSUB 140:GOTO 1680
1710 RG=(AC/60)%GS:RG=CINT(RG)
1720 GOSUB 130:PRINT"RATE OF CLIMB/DECENT: ";RG;" FPM":GOSUB 130
1730 PRINT:GOTO 170
1740 INPUT "LOW ALTITUDE (FEET)?             ",LA:PRINT
1750 INPUT "HIGH ALTITUDE (FEET)?            ",HA:PRINT
1760 GOSUB 190:IF CORRECT$<>"N" THEN 1770 ELSE GOSUB 140:GOTO 1740
1770 HH=HA-LA
1780 GOSUB 130:PRINT"ALTITUDE CHANGE: ";HH;" FEET":GOSUB 130
1790 PRINT:GOTO 170
```

Figure 34.1. The program computes a variety of flight-related data.

CHAPTER THIRTY-FIVE

Cruise Performance Data for Piston-Engine Aircraft

This program calculates true airspeed, fuel flow, endurance and range data for different power settings and altitudes for a piston-engine aircraft. Certain book data will have previously been entered. Before the program is used for the first time, the performance parameters for the aircraft must be keyed in. They are then saved in several data files and used to perform the calculations necessary to produce the desired results.

The program displays the main menu:

```
                        Do you want to:
---------------------------------------------------------------
   1  Use previously entered data.
   2  Enter new book data.
---------------------------------------------------------------
   3  Review previously entered data.
---------------------------------------------------------------
   4  Exit program.
---------------------------------------------------------------
Which?
```

The first time the program is used, the second choice must be selected. It asks:

```
Are you sure? Previous data will be erased!
```

as a reminder that data entered previously will be lost if the answer is affirmative. This is followed by a series of prompts:

```
Aircraft identification? (for instance: Cessna Skyhawk) N-number?

True airspeed (knots or mph) at 40% power at 2000 feet?
Fuel flow (gph or pph) at 40% power at 2000 feet?
```

Both the last two questions are repeated for power settings from 40 to 75 percent at five-percent increments. Once these data have been entered, they are stored in two sequential data files to be used during all subsequent runs of the program.

To examine these stored data, the third choice from the menu causes them to be displayed. Otherwise the user will select the first choice, which produces the following prompts:

```
Enter trip data for Piper Arrow N557PG
```

```
Flight altitude?
Temperature in centigrade?
Percent power?
Fuel on board (gallons or pounds)?
```

This instantly produces the appropriate data, represented here by arbitrarily chosen values:

```
True airspeed:          140.1 knots or mph
Fuel flow:              13.1 gph or pph
Endurance:              5 hours and 53 minutes
Range:                  825 nm or sm
```

The program (Figure 35.1) can be used to produce comparisons between different altitudes and power settings. These can be printed by the line printer for easier reading. (See Figure 35.2.)

Lines 100–300 display a description of the program and the main menu. Lines 310–580 key in the performance data for the aircraft for which the program is to be used. Lines 590–760 display all previously entered performance data. Lines 770–1130 determine the results based on given altitude, temperature and power-setting figures, with lines 1060–1130 performing the printing function.

```
100 REM CRUISE PERFORMANCE DATA FOR PISTON-ENGINE AIRCRAFT
110 REM BASIC-80
120 GOTO 180
130 PRINT STRING$(80,45):RETURN
140 HOME:VTAB(10):RETURN
150 PRINT:INPUT "Press >RETURN< (Q to quit)  ",R$:IF R$="Q" THEN 160 ELSE RETURN
160 GOSUB 140:GOSUB 130:PRINT TAB(38)"End.":GOSUB 130:END
170 PRINT:INPUT "Entries correct? (Y/N)    ;CORRECT$:RETURN
180 GOSUB 140:GOSUB 130
190 PRINT TAB(10)"This program determines the cruise performance data for different":PRINT
200 PRINT TAB(10)"power settings for piston-engine aircraft.":GOSUB 130
210 PRINT TAB(10)"Before using the program for the first time you must key in certain":PRINT
```

continued

```
220 PRINT TAB(10)"data from the owner's manual for your aircraft.":GOSUB 130
230 GOSUB 150:GOSUB 140
240 PRINT TAB(20)"Do you want to:":GOSUB 130
250 PRINT 1,"Use previously entered data."
260 PRINT 2,"Enter new book data.":GOSUB 130
270 PRINT 3,"Review previously entered data.":GOSUB 130
280 PRINT 4,"Exit program.":GOSUB 130
290 INPUT "Which? ",WHICH:GOSUB 140
300 ON WHICH GOTO 770,310,590,160
310 PRINT"Are you sure?  Previous data will be erased!";
320 INPUT " (Y/N)       ",YN$
330 IF YN$<>"Y" THEN GOSUB 140:GOTO 240 ELSE 340
340 OPEN "O",#1,"AIRCR.DTA"
350 INPUT "Aircraft identification?             ",AIRC$
360 PRINT #1,AIRC$
370 INPUT "N-number?                            ",NUM$
380 PRINT #1,NUM$
390 GOSUB 170:IF CORRECT$<>"N" THEN 400   ELSE GOSUB 140:GOTO 350
400 CLOSE #1:GOSUB 140
410 OPEN "O",#3,"SPEED.DTA"
420 FOR X=40 TO 75 STEP 5
430 A=A+1
440 PRINT TAB(15)"True airspeed (knots or mph) at ";X;"% power at 2000 feet":PRINT
450 PRINT:INPUT TAS(A):GOSUB 140
460 PRINT #3,TAS(A)
470 NEXT X
480 GOSUB 170:IF CORRECT$<>"N" THEN 490 ELSE GOSUB 140:GOTO 420
490 A=0:CLOSE #3:GOSUB 140
500 OPEN "O",#2,"GAS.DTA"
510 FOR X=40 TO 75 STEP 5
520 A=A+1
530 PRINT TAB(15)"Fuel flow (gph or pph) at ";X;"% power at 2000 feet":PRINT
540 PRINT:INPUT FUEL(A):GOSUB 140
550 PRINT #2,FUEL(A)
560 NEXT X
570 GOSUB 170:IF CORRECT$<>"N" THEN 580 ELSE GOSUB 140:GOTO 510
580 A=0:CLOSE #2:GOSUB 140:GOTO 240
590 OPEN "I",#1,"AIRCR.DTA"
600 INPUT #1,AIRC$,NUM$
610 PRINT"Performance data for ";AIRC$;" ";NUM$
620 CLOSE #1:GOSUB 130
630 OPEN "I",#3,"SPEED.DTA"
640 FOR X=40 TO 75 STEP 5
650 A=A+1
660 INPUT #3,TAS(A)
670 PRINT"True airspeed at ";X;"% power= ";TAS(A);" knots or mph.":PRINT
680 NEXT X
690 A=0:CLOSE #3:GOSUB 150:GOSUB 140
700 OPEN "I",#2,"GAS.DTA"
710 FOR X=40 TO 75 STEP 5
720 A=A+1
730 INPUT #2,FUEL(A)
740 PRINT"Fuel flow at ";X;"% power= ";FUEL(A);" gph or pph.":PRINT
750 NEXT X
760 A=0:CLOSE #2:GOSUB 150:GOSUB 140:GOTO 240
770 OPEN "I",#1,"AIRCR.DTA"
780 A=A+1
790 INPUT #1,AIRC$,NUM$
800 PRINT"Enter trip data for ";AIRC$;" ";NUM$:GOSUB 130
810 Q(1)=.001:Q(2)=.0007:Q(3)=.00065:Q(4)=.0009:Q(5)=.00105
820 Q(6)=.0012:Q(7)=.00125:Q(8)=.00085
830 INPUT "Flight altitude?                  ",FA:PRINT
840 INPUT "Temperature in centigrade?        ",CC:PRINT
850 INPUT "Percent of power?                 ",PP:PRINT
```

```
860 INPUT "Fuel on board (gallons or pounds)?   ",FB:PRINT
870 GOSUB 170:IF CORRECT$<>"N" THEN 880 ELSE GOSUB 140:GOTO 830
880 CLOSE #1:OPEN "I",#2,"GAS.DTA"
890 OPEN "I",#3,"SPEED.DTA"
900 FOR X=40 TO 75 STEP 5
910 INPUT #3,TAS(A)
920 INPUT #2,FUEL(A)
930 B=B+1:Q=Q(B)
940 IF X=PP THEN 950 ELSE NEXT X
950 ZZ=(FA%Q)+TAS(A):II=FUEL(A)
960 QQ=II%.75:QQ=FB-QQ
970 TT=QQ/II:TT=INT(TT%100+.5)/100
980 SS=INT(TT):MM=TT-SS:MM=MM%.6:MM=MM%100:MM=INT(MM+.5)
990 OO=ZZ%TT:OO=INT(OO+.5):GOSUB 140
1000 PRINT"True airspeed:            ";ZZ;" knots or mph":PRINT
1010 PRINT"Fuel flow:               ";II;" gph or pph":PRINT
1020 PRINT"Endurance:               ";SS;" hours and ";MM;" minutes":PRINT
1030 PRINT"Range:                   ";OO;" nm or sm":GOSUB 130
1040 INPUT "Printout?  (Y/N)   ",PR$:IF PR$<>"Y" THEN 1050 ELSE 1060
1050 A=0:B=0:CLOSE #3:CLOSE #2:GOSUB 150:GOSUB 140:GOTO 240
1060 LPRINT AIRC$;" ";NUM$:LPRINT
1070 LPRINT "Altitude:        ";FA:LPRINT
1080 LPRINT "Percent power:   ";PP:LPRINT STRING$(20,61)
1090 LPRINT "TAS";TAB(20)"FUEL FLOW";TAB(40)"ENDURANCE";TAB(60)"RANGE"
1100 LPRINT STRING$(18,45);TAB(20)STRING$(18,45);TAB(40)STRING$(18,45);
1110 LPRINT TAB(60)STRING$(18,45)
1120 LPRINT ZZ;TAB(20)II;TAB(40)SS;":";MM;TAB(60)OO
1130 GOTO 1050
```

Figure 35.1   The cruise-performance program for piston-engine aircraft.

```
PIPER ARROW N557PG

Altitude:         9000

Percent power:    55
=====================
TAS                FUEL FLOW            ENDURANCE            RANGE
------------------ -------------------  -------------------  -------------------
 140.1              13.1                 5 : 53               825
```

Figure 35.2   A sample print-out produced by the cruise performance program.

CHAPTER THIRTY-SIX

Flight Route File Program

Most pilots find they fly the same routes, time after time. It would be convenient if the appropriate data, such as navigation aid identifiers, frequencies, magnetic bearings from one route segment to the next and so on, were available with an estimate of the time en route, based on forecast winds aloft. The amount of fuel needed would also be handy.

The program reproduced here performs that function with minimum effort by the user. The user must only key in the data for the different route segments one time. Once they have been entered, the program creates two data files in which the information is stored and always available for future recall.

Any time a flight over these route segments is scheduled, the pilot need only key in the wind, cruise speed and fuel flow data, and the program does the rest. All necessary prompts have been built into the program, and the pilot need not remember the order in which data are entered. It works like this:

When the program is activated, it displays an explanation of its purpose and then confronts the user with a menu of six choices:

```
Menu:
----------------------------------------------------------------------
1  Use previously stored routes.
2  Examine previously stored routes by route number.
3  Examine all stored routes.
4  Examine last used route number.
5  Enter new route data.
----------------------------------------------------------------------
6  Exit program.
----------------------------------------------------------------------
Which?
```

The first time the program is used, the pilot must select choice number five and enter the route segment data as follows:

```
Route or route segment number?        (to 3 digits)
Airport or fix identifier?            (3 letters)
Frequency (VOR)?                      (xxx.xxx)
Latitude?                             (xx-xx.xN)
Longitude?                            (xxx-xx.xW)
Victor airway?                        (Vxxx)
Distance to next fix (nm)             (to 4 digits)
Magnetic course                       (to 3 digits)
Magnetic variation (E-/W+)            (+/-2 digits)
```

After the pilot has keyed in the information, the program asks whether another segment is to be entered. At this stage the pilot should enter all route segment data for one or more routes flown with frequency. Additional routes and route segments can be added at any later date. The individual segments must be numbered consecutively (from 1–999) and the individual segments, making up a route, must be entered in consecutive order. In other words, segments 1 through 8 might be a route from Santa Fe to Los Angeles, segments 9 through 15 might be a route from Los Angeles to San Francisco, and so on. If the pilot does not remember the last used route number, he can select choice number four from the main menu. The computer will display:

```
The last used route number was 14. Next use 15.
```

When the program is used prior to a scheduled flight, the pilot selects choice number one. The display then again asks for certain input:

```
First route segment number to be used?
Last route segment number to be used?
Cruise speed?    (knots)
Fuel flow?    (gph or pph)
Wind direction?    (degrees)
Wind velocity?    (knots)
```

```
Route segment #  1
------------------------------------------
VOR:                  SAF
Frequency=            110.6
Lat/Long=             35-32.4N / 106-03.9W
Victor airway=        V19
Distance to next fix  49    nm
Magnetic course=      232 degrees
Magnetic variation=   -12
------------------------------------------

Route segment #  2
------------------------------------------
VOR:                  ABQ
Frequency=            113.2
Lat/Long=             35-02.6N / 106-49.0W
Victor airway=        V190
Distance to next fix  121   nm
Magnetic course=      240 degrees
Magnetic variation=   -12
------------------------------------------

Route segment #  3
------------------------------------------
VOR:                  SJN
Frequency=            112.3
Lat/Long=             34-25.4N / 109-08.6W
Victor airway=        V190
Distance to next fix  149   nm
Magnetic course=      235 degrees
Magnetic variation=   -11
------------------------------------------
```

Figure 36.1 A sample print-out produced by the flight route file program.

The program uses these inputs along with the previously stored data to produce a display that looks something like this:

Route segment #   5

Time en route:	0 hours and 42 minutes
Fuel used:	11.7 gallons or pounds
Total time to this point:	3 hours and 17 minutes
Total fuel burn to this point:	44.7 gallons or pounds

This display is repeated consecutively for each route segment and the data may, at the user's option, be sent to the line printer (see Figure 36.1).

The program (Figure 36.2) consists of 135 lines. Lines 100–400 are used to display a description of the program and the main menu, plus performing the assignment of various string variables. Lines 410–770 are used if previously recorded data are to be used to determine the flight parameters under current wind conditions. Lines 780–900 are used if the pilot wants to look at a given route segment. Lines 910–940 are used if the pilot wants to see the last used route number. Lines 950–1160 are used if new route segment data are to be entered. These data are recorded in two data files, a sequential file called RTE.NUM which stores the last used route segment number, and a random-access file called ROUTE.DTA which stores the data for each route segment. Finally, lines 1170–1350 are used if the pilot wants to take a look at all previously stored data, one route segment after another.

```
100 REM FLIGHT ROUTE FILE PROGRAM
110 REM BASIC-80
120 GOTO 180
130 PRINT STRING$(80,45):RETURN
140 HOME:VTAB(10):RETURN
150 PRINT:INPUT "Press >RETURN< (Q to quit)    ",R$:IF R$="Q" THEN 160 ELSE RETURN
160 GOSUB 140:GOSUB 130:PRINT TAB(38)"End.":GOSUB 130:END
170 PRINT:INPUT "Entry correct?  (Y/N)   ",CORRECT$:RETURN
180 ZZ=0:GOSUB 140:GOSUB 130
190 PRINT TAB(7)"This program stores the data for frequently used routes, computing":PRINT
200 PRINT TAB(7)"the time en route and fuel consumption based on input wind data.":GOSUB 130
210 PRINT TAB(7)"You may store as many routes and route segments as you like.":GOSUB 130
220 GOSUB 150:GOSUB 140
230 PRINT TAB(10)"Menu:":GOSUB 130
240 PRINT 1,"Use previously stored routes."
250 PRINT 2,"Examine previously stored routes by route number."
260 PRINT 3,"Examine all stored routes."
270 PRINT 4,"Examine last used route number."
280 PRINT 5,"Enter new route data.":GOSUB 130
```

continued

```
290 PRINT 6,"Exit program.":GOSUB 130
300 INPUT "Which?    ",WHICH:GOSUB 140
310 VV$="VOR:                   "
320 FF$="Frequency:             "
330 LL$="Lat/Long:              "
340 AA$="Victor airway:         "
350 CC$="Magnetic course:       "
360 MV$="Magnetic variation:    "
370 TT$="Time en route:         "
380 DD$="Distance to next fix:  "
390 LLL$="-------------------------------------------------"
400 ON WHICH GOTO 410,780,1170,910,950,160
410 OPEN "R",#1,"ROUTE.DTA",50
420 FIELD #1,3 AS A$,3 AS B$,7 AS C$,8 AS D$,9 AS E$,5 AS F$,4 AS G$,3 AS H$,3 AS I$
430 INPUT "First route segment number to be used?          ",ROUTE1%
440 INPUT "Last route segment number to be used?           ",ROUTE2%
450 FOR XX=ROUTE1% TO ROUTE2%
460 GET #1,XX
470 IF ZZ=5 THEN 530
480 INPUT "Cruise speed?     (knots)               ",KNOTS
490 INPUT "Fuel flow?        (gph or pph)          ",FUEL
500 INPUT "Wind direction?   (degrees)             ",WD
510 INPUT "Wind velocity?    (knots)               ",WV:GOSUB 130
520 GOSUB 170:IF CORRECT$<>"N" THEN 530 ELSE GOSUB 140:GOTO 480
530 PRINT"Route segment #";XX:GOSUB 130
540 PRINT VV$;B$:PRINT FF$;C$:PRINT LL$;D$;" / ";E$:PRINT AA$;F$
550 PRINT DD$;G$;" nm":PRINT CC$;H$;" degrees":PRINT MV$;I$:GOSUB 130
560 WC=-1*WV*COS((WD-VAL(H$)-VAL(I$))/57.3)
570 CC=CC+WC:TT=VAL(G$)/KNOTS:TF=TT:T=INT(TT):TT=TT-T
580 CC=INT(CC):TT=TT*.6*100:TT=INT(TT+.5)
590 FB=TF*FUEL:FB=INT(FB*10+.5)/10
600 PRINT TT$;"         ";T;" hours and ";TT;" minutes.":PRINT
610 PRINT"Fuel used:                      ";FB;" gallons or pounds.":GOSUB 130
620 TTF=TTF+TF:TFF1=INT(TTF):TTF1=TTF-TFF1:TTF1=TTF1*.6*100:TTF1=INT(TTF1+.5)
630 FFB=FFB+FB
640 PRINT"Total time to this point:       ";TFF1;" hours and ";TTF1;" minutes.":PRINT
650 PRINT"Total fuel burn to this point: ";FFB;" gallons or pounds."
660 ZZ=5:GOSUB 130:INPUT "Printout?  (Y/N)          ",RP$
670 IF RP$="Y" THEN 700 ELSE 680
680 GOSUB 140:NEXT XX
690 GOSUB 140:GOTO 230
700 LPRINT "Route segment #";XX:LPRINT LLL$
710 LPRINT VV$;B$:LPRINT FF$;C$:LPRINT LL$;D$;" / ";E$:LPRINT AA$;F$
720 LPRINT DD$;G$;" nm":LPRINT CC$;H$;" degrees":LPRINT MV$;I$:LPRINT LLL$
730 LPRINT TT$;"         ";T;" hours and ";TT;" minutes."
740 LPRINT "Fuel used:                      ";FB;" gallons or pounds.":LPRINT LLL$
750 LPRINT "Total time to this point:       ";TFF1;" hours and ";TTF1;" minutes."
760 LPRINT "Total fuel burn to this point: ";FFB;" gallons or pounds."
770 LPRINT LLL$:LPRINT:GOTO 680
780 INPUT "Route number to be examined?            ",ROUTE%
790 GOSUB 140:OPEN "R",#1,"ROUTE.DTA",50
800 FIELD #1,3 AS A$,3 AS B$,7 AS C$,8 AS D$,9 AS E$,5 AS F$,4 AS G$,3 AS H$,3 AS I$
810 GET #1,ROUTE%
820 PRINT"Route #";CVI(A$):GOSUB 130
830 PRINT VV$;B$:PRINT FF$;C$:PRINT LL$;D$;" / ";E$:PRINT AA$;F$
835 PRINT DD$;G$;" nm":PRINT CC$;H$;" degrees":PRINT MV$;I$
840 GOSUB 130:INPUT "Printout?   (Y/N)                       ",PR$
850 IF PR$<>"Y" THEN 900 ELSE 860
860 LPRINT "Route #";CVI(A$):LPRINT LLL$
870 LPRINT VV$;B$:LPRINT FF$;C$:LPRINT LL$;D$;" / ";E$
880 LPRINT AA$;F$:LPRINT DD$;G$;" nm":LPRINT CC$;H$;" degrees":LPRINT MV$;I$
890 LPRINT LLL$:LPRINT
900 CLOSE #1:GOSUB 140:GOTO 230
910 OPEN "I",#2,"RTE.NUM"
```

```
920 INPUT #2,ROUTEZ
930 PRINT"The last used route number was ";ROUTEZ;". Next use ";ROUTEZ+1
940 CLOSE #2:GOSUB 150:GOSUB 140:GOTO 230
950 OPEN "R",#1,"ROUTE.DTA",50
960 FIELD #1,3 AS A$,3 AS B$,7 AS C$,8 AS D$,9 AS E$,5 AS F$,4 AS G$,3 AS H$,3 AS I$
970 INPUT "Route or route segment number? (to 3 digits)    ",ROUTEZ
980 INPUT "Airport or fix identifier?     (3 letters)     ",AIR$
990 INPUT "Frequency (VOR)?               (xxx.xxx)        ",FREQ$
1000 INPUT "Latitude?                     (xx-xx.xN)       ",LAT$
1010 INPUT "Longitude?                    (xxx-xx.xW)      ",LONG$
1020 INPUT "Victor airway?                (Vxxx)           ",VICTOR$
1030 INPUT "Distance to next fix (nm)     (to 4 digits)    ",NM$
1040 INPUT "Magnetic course               (to 3 digits)    ",COURSE$
1050 INPUT "Magnetic variation (E-/W+)    (+/- 2 digits)   ",VAR$
1060 GOSUB 170:IF CORRECT$<>"N" THEN 1070 ELSE GOSUB 140:GOTO 970
1070 OPEN "O",#2,"RTE.NUM"
1080 PRINT #2,ROUTEZ
1090 CLOSE #2
1100 LSET A$=MKI$(ROUTEZ)
1110 LSET B$=AIR$:LSET C$=FREQ$:LSET D$=LAT$:LSET E$=LONG$
1120 LSET F$=VICTOR$:LSET G$=NM$:LSET H$=COURSE$:LSET I$=VAR$
1130 PUT #1,ROUTEZ
1140 GOSUB 130:INPUT "Enter another route?                        ",YN$
1150 IF YN$="Y" THEN 970 ELSE 1160
1160 CLOSE #1:GOSUB 140:GOTO 230
1170 OPEN "I",#2,"RTE.NUM"
1180 INPUT #2,ROUTEZ
1190 CLOSE #2
1200 OPEN "R",#1,"ROUTE.DTA",50
1210 FIELD #1,3 AS A$,3 AS B$,7 AS C$,8 AS D$,9 AS E$,5 AS F$,4 AS G$,3 AS H$,3 AS I$
1220 INPUT "Printout?  (Y/N)     ",PP$
1230 IF PP$<>"Y" THEN 1300 ELSE 1240
1240 FOR XXX=1 TO ROUTEZ:GET #1,XXX
1250 LPRINT "Route segment # ";XXX:LPRINT LLL$
1260 LPRINT VV$;B$:LPRINT FF$;C$:LPRINT LL$;D$;" / ";E$
1270 LPRINT AA$;F$:LPRINT DD$;G$;" nm":LPRINT CC$;H$;" degrees"
1280 LPRINT MV$;I$:LPRINT LLL$:LPRINT:NEXT XXX
1290 GOTO 1350
1300 FOR XXX=1 TO ROUTEZ:GET #1,XXX
1310 PRINT"Route segment # ";XXX:GOSUB 130
1320 PRINT VV$;B$:PRINT FF$;C$:PRINT LL$;D$;" / ";E$
1330 PRINT AA$;F$:PRINT DD$;G$;" nm":PRINT CC$;H$;" degrees"
1340 PRINT MV$;I$:GOSUB 130:GOSUB 150:NEXT XXX
1350 CLOSE #1:GOSUB 150:GOSUB 140:GOTO 230
```

Figure 36.2 The flight route file program.

CHAPTER THIRTY-SEVEN

Great Circle Navigation

Though it may seem illogical, on long trips the shallow curves represented by the Great Circle Routes are shorter than what appears on the charts as a straight line. While most aircraft that fly such distances are equipped with avionics systems that automatically compute the Great Circle Routes between two points, this program can be used by pilots flying aircraft not equipped with fancy and expensive black boxes.

The program uses the latitude and longitude data for the departure and destination points, and produces a display or option print-out of as many legs of that route as desired. Figure 37.1 is such a print-out for a flight from LGA (LaGuardia, New York) to LAX (Los Angeles International). The departure and destination point data were entered, rounded to the nearest minute, along with the magnetic variation for the first leg of the trip. Intermediate longitudes were entered in five-degree intervals, and each time the computer displayed and printed the intermediate latitude, the remaining distance to the destination, the true course for the next leg, and the leg distance for the leg just flown.

```
Departure point latitude:              40.47
Departure point longitude:             73.53
Destination, latitude:                 33.56
Destination, longitude:                118.26
------------------------------------------------------
Departure, mag variation:              12
------------------------------------------------------

Great Circle distance to destination:  2159  nautical miles.
------------------------------------------------------
The true course:                       274  degrees.
------------------------------------------------------
The magnetic course is                 286  degrees
------------------------------------------------------
```

```
Intermediate longitude:                    78
--------------------------------------------------------
The intermediate latitude is               40   degrees
--------------------------------------------------------
Great Circle distance to destination:      1956  nautical miles.
--------------------------------------------------------
The true course:                           272  degrees.
--------------------------------------------------------
The leg distance is                        203  nautical miles
--------------------------------------------------------

Intermediate longitude:                    83
--------------------------------------------------------
The intermediate latitude is               39   degrees
--------------------------------------------------------
Great Circle distance to destination:      1725  nautical miles.
--------------------------------------------------------
The true course:                           270  degrees.
--------------------------------------------------------
The leg distance is                        231  nautical miles
--------------------------------------------------------

Intermediate longitude:                    88
--------------------------------------------------------
The intermediate latitude is               38   degrees
--------------------------------------------------------
Great Circle distance to destination:      1490  nautical miles.
--------------------------------------------------------
The true course:                           269  degrees.
--------------------------------------------------------
The leg distance is                        235  nautical miles
--------------------------------------------------------

Intermediate longitude:                    93
--------------------------------------------------------
The intermediate latitude is               37   degrees
--------------------------------------------------------
Great Circle distance to destination:      1251  nautical miles.
--------------------------------------------------------
The true course:                           268  degrees.
--------------------------------------------------------
The leg distance is                        239  nautical miles
--------------------------------------------------------

Intermediate longitude:                    98
--------------------------------------------------------
The intermediate latitude is               36   degrees
--------------------------------------------------------
Great Circle distance to destination:      1007  nautical miles.
--------------------------------------------------------
The true course:                           268  degrees.
--------------------------------------------------------
The leg distance is                        244  nautical miles
--------------------------------------------------------

Intermediate longitude:                    103
--------------------------------------------------------
The intermediate latitude is               35   degrees
--------------------------------------------------------
Great Circle distance to destination:      761  nautical miles.
```

continued

```
-----------------------------------------------------
The true course:                    268  degrees.
-----------------------------------------------------
The leg distance is                 246  nautical miles
-----------------------------------------------------

Intermediate longitude:             108
-----------------------------------------------------
The intermediate latitude is        34  degrees
-----------------------------------------------------
Great Circle distance to destination:  512  nautical miles.
-----------------------------------------------------
The true course:                    270  degrees.
-----------------------------------------------------
The leg distance is                 249  nautical miles
-----------------------------------------------------

Intermediate longitude:             113
-----------------------------------------------------
The intermediate latitude is        33  degrees
-----------------------------------------------------
Great Circle distance to destination:  266  nautical miles.
-----------------------------------------------------
The true course:                    279  degrees.
-----------------------------------------------------
The leg distance is                 246  nautical miles
-----------------------------------------------------

Intermediate longitude:             118
-----------------------------------------------------
The intermediate latitude is        33  degrees
-----------------------------------------------------
Great Circle distance to destination:  36  nautical miles.
-----------------------------------------------------
The true course:                    339  degrees.
-----------------------------------------------------
The leg distance is                 230  nautical miles
-----------------------------------------------------
```

Figure 37.1 A sample print-out produced by the great circle navigation program.

When the program is called up, it asks for the initial input:

Great Circle Navigation

```
-----------------------------------------------------------------
Departure point latitude?            (+N –S)
Departure point longitude?           (+W –E)
Destination, latitude?               (+N –S)
Destination, longitude?              (+W –E)
-----------------------------------------------------------------
Departure, mag variation?            (+W –E)
-----------------------------------------------------------------
Printout? (Y/N)
```

After this it simply repeats over and over:

Intermediate longitude?

and the computer does the rest.

Program (Figure 37.2) lines 100–300 assign a value to PI (line 120), contain a number of subroutines, and define several functions in lines 200 and 210. Then the input prompts are displayed. Line 310 is a branch used later. Lines 320–380 cause the initial input to be sent to the line printer if the print-out option is in effect. Lines 390–950 perform the large number of computations needed to consecutively display and/or print the intermediate results.

```
100 REM GREAT CIRCLE NAVIGATION
110 REM BASIC-80
120 PI=3.14159:GOTO 200
130 PRINT STRING$(80,45):RETURN
140 LPRINT STRING$(50,45):RETURN
150 HOME:VTAB(10):RETURN
160 PRINT:INPUT "Press >RETURN< (Q to quit)     ",R$
170 IF R$="Q" THEN 180 ELSE RETURN
180 GOSUB 150:GOSUB 130:PRINT TAB(38)"End.":GOSUB 130:END
190 PRINT:INPUT "Entry correct?  (Y/N)  ",CORRECT$:RETURN
200 DEF FNACOS(A)=-ATN(A/SQR(-A*A+1))+1.5708
210 SF=0:DEF FNRAD(A)=A*(PI/180):DEF FNDEG(A)=A/(PI/180)
220 GOSUB 150:GOSUB 130
230 PRINT TAB(20)"Great Circle Navigation":GOSUB 130:GOSUB 160:GOSUB 150
240 INPUT "Departure point latitude?     (+N  -S)   ",A1:A28=A1
250 INPUT "Departure point longitude?    (+W  -E)   ",A2:A27=A2
260 INPUT "Destination, latitude?        (+N  -S)   ",A3:A29=A3
270 INPUT "Destination, longitude?       (+W  -E)   ",A4:A26=A4:GOSUB 130
280 INPUT "Departure, mag variation?     (+W  -E)   ",A19
290 GOSUB 190:IF CORRECT$<>"N" THEN 300 ELSE GOSUB 150:GOTO 240
300 GOSUB 130:INPUT "Printout?  (Y/N)  ",PR$
310 GOTO 330
320 A40=FNDEG(A40):A1=A39:A27=A40:A2=A40:A3=A29:A4=A26:GOTO 390
330 IF PR$<>"Y" THEN 390 ELSE 340
340 LPRINT "Departure point latitude:          ",A1
350 LPRINT "Departure point longitude:         ",A2
360 LPRINT "Destination, latitude:             ",A3
370 LPRINT "Destination, longitude:            ",A4:GOSUB 140
380 LPRINT "Departure, mag variation:          ",A19:GOSUB 140:LPRINT
390 A5=COS(FNRAD(A3)-FNRAD(A1))
400 A8=FNACOS((COS(FNRAD(A4)-FNRAD(A2))-1)*COS(FNRAD(A1))*COS(FNRAD(A3))+A5)
410 A0=A8*60:A0=FNDEG(A0):A0=CINT(A0):GOSUB 150:GOSUB 130
420 PRINT"Great Circle distance to destination:    ";A0;" nautical miles.":GOSUB 130
430 IF PR$<>"Y" THEN 450 ELSE 440
440 LPRINT "Great Circle distance to destination:    ";A0;" nautical miles.":GOSUB 140
450 SWAP A1,A3:A2=360:A4=A27
460 A9=SIN(FNRAD(A1)):A10=SIN(A8):A11=COS(FNRAD(A3)):A12=TAN(FNRAD(A3))
470 A31=TAN(FNRAD(A1))
480 A13=A9/A10/A11:A14=A12/TAN(A8):A6=FNACOS(A13-A14)
490 A15=SIN(FNRAD(A2)-FNRAD(A4))
500 A6=FNDEG(A6)
510 IF 0<=A15 THEN 680 ELSE 520
520 A6=A6-360:IF A6<0 THEN A6=A6-(A6*2)
530 A6=CINT(A6)
```

continued

```
540 PRINT"The true course:                              ";A6;" degrees.":GOSUB 130
550 IF PR$<>"Y" THEN 570 ELSE 560
560 LPRINT "The true course:                            ";A6;" degrees.":GOSUB 140
570 IF SF=2 THEN 580 ELSE 630
580 A42=ABS(A0-A41):A42=CINT(A42)
590 PRINT"The leg distance is                           ";A42;" nautical miles":GOSUB 130
600 IF PR$<>"Y" THEN 620 ELSE 610
610 LPRINT "The leg distance is                         ";A42;" nautical miles":GOSUB 140
620 GOTO 900
630 A19=A6+A19
640 PRINT"The magnetic course is                        ";A19;" degrees":GOSUB 130
650 IF PR$<>"Y" THEN 670 ELSE 660
660 LPRINT "The magnetic course is                      ";A19;" degrees":GOSUB 140
670 IF SF<>2 THEN 900 ELSE 820
680 A6=CINT(A6*.6):GOTO 540
690 IF A4<=0 THEN A4=A4+360
700 IF A2>A4 THEN A22=A2
710 A24=A4:A23=A1:A25=A3:A4=A22:A2=A24:A3=A23:A1=A25
720 A30=SIN(A7-FNRAD(A26)):AA31=A26-A24:AA32=FNDEG(A7)-A24
730 A31=SIN(FNRAD(AA31)):A32=SIN(FNRAD(AA32))*A21
740 A33=A30*A12:A18=ATN((FNRAD(A32)-FNRAD(A33))/FNRAD(A31))
750 A18=INT(FNDEG(A18))
760 IF SF=2 THEN GOSUB 130:GOTO 770 ELSE 900
770 PRINT"The intermediate latitude is                  ";A18;" degrees":GOSUB 130:GOSUB 160
780 IF PR$<>"Y" THEN 800 ELSE 790
790 LPRINT "The intermediate latitude is                ";A18;" degrees":GOSUB 140
800 IF SF=2 THEN A39=A18
810 GOTO 320
820 A14=TAN(FNRAD(A6))*SIN(FNRAD(A1)):A15=ATN(1/A14):A15=(AA7-A15)*.6
830 IF 0>A15 THEN 850
840 IF 180<=A15 THEN 860 ELSE 870
850 A15=A15+360:GOTO 870
860 A15=A15-360:IF A15<0 THEN A15=A15-(A15*2)
870 A7=A15:A4=A24/.6
880 PRINT"14 ";A14;" 15 ";A15;" 7 ";A7;" 6 ";A6;" 1 ";A1;" 24 ";A24
890 GOTO 690
900 INPUT "Intermediate longitude?                      ",A7:AA7=A7
910 GOSUB 190:IF CORRECT$<>"N" THEN 920 ELSE GOSUB 150:GOTO 900
920 IF PR$<>"Y" THEN 940 ELSE 930
930 LPRINT:LPRINT "Intermediate longitude:              ",A7:GOSUB 140
940 A7=FNRAD(A7):A40=A7:A27=A7
950 A41=A0:SF=2:GOTO 690
```

Figure 37.2   The great circle navigation program.

CHAPTER THIRTY-EIGHT

Direct Route Time and Fuel Data

This program determines the distance between two points referenced to one VOR. It then uses wind direction and velocity data to compute the time en route and the amount of fuel required in both directions (Point A to Point B and Point B to Point A).

When the program is run, it starts by asking that a series of data be keyed in:

Point A, distance from VOR in nm?
Point A, radial from VOR?

Point B, distance from VOR in nm?
Point B, radial from VOR?

Wind direction at cruise level?
Wind velocity at cruise level in knots?

True air speed at cruise level in knots?

Magnetic course, point A to point B?
Average magnetic variation? (E −/W +)

Average fuel flow in gph or pph?

Based on these inputs, the computer performs a long series of calculations to arrive at the following answers (here using arbitrary figures):

The distance between points A and B is 172 nm.

At a TAS of 175 knots, your ground speed
from point A to point B will be 158 knots.

At a TAS of 175 knots, your ground speed
from point B to point A will be 186 knots.

The time en route from point A to point B
will be 1 hour and 5 minutes.

The time en route from point B to point A
will be 0 hours and 55 minutes.

The fuel required from point A to point B
will be 27 gallons or pounds.

The fuel required from point B to point A
will be 23 gallons or pounds.

Another run? (Y/N)

```
100 REM DIRECT ROUTE TIME & FUEL DATA
110 REM BASIC-80
120 GOTO 200
130 PRINT STRING$(80,45):RETURN
140 LPRINT STRING$(50,45):RETURN
150 HOME:VTAB(10):RETURN
160 PRINT:INPUT "Press >RETURN<  (Q to quit)  ",R$
170 IF R$="Q" THEN 180 ELSE RETURN
180 GOSUB 150:GOSUB 130:PRINT TAB(38)"End.":GOSUB 130:END
190 PRINT:INPUT "Entries correct?  (Y/N)  ",CORRECT$:RETURN
200 GOSUB 150
210 PRINT TAB(25)"DIRECT ROUTE TIME & FUEL DATA"
220 PRINT TAB(25)STRING$(29,42):PRINT
230 PRINT"This program determines the distance between two points referenced to"
240 PRINT"one VOR, the time en route in either direction and the fuel required"
250 PRINT"based on forecast or actual winds aloft.":GOSUB 130:GOSUB 160:GOSUB 150
260 INPUT "Point A, distance from VOR in nm?      ",NM1
270 INPUT "Point A, radial from VOR?              ",RAD1:GOSUB 130
280 INPUT "Point B, distance from VOR in nm?      ",NM2
290 INPUT "Point B, radial from VOR?              ",RAD2:GOSUB 130
300 INPUT "Wind direction at cruise level?        ",WD
310 INPUT "Wind velocity at cruise level in knots? ",WV:GOSUB 130
320 INPUT "True airspeed at cruise level in knots? ",TAS:GOSUB 130
330 INPUT "Magnetic course, point A to point B?   ",MC
340 INPUT "Average magnetic variation? (E -/ W +) ",MV:GOSUB 130
```

```
350 INPUT "Average fuel flow in gph or pph?            ",FUEL
360 GOSUB 190:IF CORRECT$<>"N" THEN 370 ELSE GOSUB 150:GOTO 260
370 GOSUB 130:GOSUB 150
380 IF RAD1-RAD2>180 OR RAD2-RAD1>180 THEN 390 ELSE 470
390 IF RAD1>=270 AND RAD1<=360 THEN 400 ELSE 410
400 RAD1=RAD1-90
410 IF RAC1>=1 AND RAD1<=90 THEN 420 ELSE 430
420 RAD1=RAD1+90
430 IF RAD2>=270 AND RAD2<=360 THEN 440 ELSE 450
440 RAD2=RAD2-90
450 IF RAD2<=90 AND RAD2>=1 THEN 460 ELSE 470
460 RAD2=RAD2+90
470 ANG=ABS(RAD1-RAD2):DIST=1/3/21.6
480 DIST1=NM1*DIST*ANG:DIST2=NM2*DIST*ANG:MILES=CINT((DIST1+DIST2)/2)
490 PRINT"The distance between points A and B is  ";MILES;" nm."
500 RAD=57.2958:DEF FNWC(WV,WD,MC,MV)=-1*WV*COS((WD-MC-MV)/RAD)
510 WC=FNWC(WV,WD,MC,MV):GS1=CINT(TAS+WC):GS2=CINT(TAS-WC):GOSUB 130
520 PRINT"At a TAS of ";TAS;" knots, your ground speed "
530 PRINT"from point A to point B will be ";GS1;" knots.":GOSUB 130
540 PRINT"At a TAS of ";TAS;" knots, your ground speed "
550 PRINT"from point B to point A will be ";GS2;" knots.":GOSUB 130
560 TIME1=MILES/GS1:TIME2=MILES/GS2
570 FUEL1=CINT(FUEL*TIME1):FUEL2=CINT(FUEL*TIME2)
580 HR1=INT(TIME1):MIN1=CINT((TIME1-HR1)*.6*100)
590 HR2=INT(TIME2):MIN2=CINT((TIME2-HR2)*.6*100)
600 PRINT"The time en route from point A to point B"
610 PRINT"will be ";HR1;" hours and ";MIN1;" minutes":GOSUB 130
620 PRINT"The time en route from point B to point A"
630 PRINT"will be ";HR2;" hours and ";MIN2;" minutes":GOSUB 130
640 PRINT"The fuel required from point A to point B"
650 PRINT"will be ";FUEL1;" gallons or pounds":GOSUB 130
660 PRINT"The fuel required from point B to point A"
670 PRINT"will be ";FUEL2;" gallons or pounds":GOSUB 130
680 INPUT "Printout?  (Y/N)   ",PR$:IF PR$<>"Y" THEN 690 ELSE 710
690 INPUT "Another run?  (Y/N)  ",YN$
700 IF YN$<>"Y" THEN 180 ELSE GOSUB 150:GOTO 260
710 LPRINT "The distance between points A and B is ";MILES;" nm."
720 GOSUB 140
730 LPRINT "At a TAS of ";TAS;" knots, your ground speed"
740 LPRINT "from point A to point B will be ";GS1;" knots."
750 GOSUB 140
760 LPRINT "At a TAS of ";TAS;" knots, your ground speed"
770 LPRINT "from point B to point A will be ";GS2;" knots."
780 GOSUB 140
790 LPRINT "The time en route from point A to point B"
800 LPRINT "will be ";HR1;" hours and ";MIN1;" minutes."
810 GOSUB 140
820 LPRINT "The time en route from point B to point A"
830 LPRINT "will be ";HR2;" hours and ";MIN2;" minutes."
840 GOSUB 140
850 LPRINT "The fuel required from point A to point B"
860 LPRINT "will be ";FUEL1;" gallons or pounds."
870 GOSUB 140
880 LPRINT "The fuel required from point B to point A"
890 LPRINT "will be ";FUEL2;" gallons or pounds."
900 GOSUB 140:LPRINT:GOTO 690
```

Figure 38.1 The direct route time and fuel data program.

Lines 100–370 of the program (Figure 38.1) display a description of the program, followed by the series of input prompts. Lines 380–480 perform the computations that produce the distance between the two

points. Lines 490–700 display the results while performing a number of
additional computations for time en route and fuel requirements.

Lines 710–900 send the results to the line printer. (Figure 38.2).

```
The distance between points A and B is   350   nm.
----------------------------------------------------
At a TAS of   145   knots, your ground speed
from point A to point B will be   122   knots.
----------------------------------------------------
At a TAS of   145   knots, your ground speed
from point B to point A will be   168   knots.
----------------------------------------------------
The time en route from point A to point B
will be   2   hours and   52   minutes.
----------------------------------------------------
The time en route from point B to point A
will be   2   hours and   5   minutes.
----------------------------------------------------
The fuel required from point A to point B
will be   36   gallons or pounds.
----------------------------------------------------
The fuel required from point B to point A
will be   26   gallons or pounds.
----------------------------------------------------
```

Figure 38.2 A sample print-out produced by the print option.

CHAPTER THIRTY-NINE

Density Altitude

I remember the first time I realized the devastating effect of density altitude. I had stopped for fuel and a leisurely lunch on a hot summer day in Albuquerque, New Mexico. I was flying a Cessna 175 at the time, a much maligned aircraft that was much better than its reputation. By two or two-thirty the thermometer was in the mid 90's, and I had to open both doors of the airplane to cool it down enough for me to get in without being boiled alive.

Eventually, I was ready to continue my flight, and ground control gave me runway 21 for take-off. That runway is 7,750 feet long, certainly long enough for someone flying a little airplane with no worry. At least, that's what I thought. After being cleared for take-off, I applied full throttle and started my run. I couldn't imagine why the airplane seemed to be taking its own sweet time to accelerate. To make a long story short, I must have used up a good two-thirds of that runway before reaching flying speed, and even then the airplane just barely hung in the air and I was darn glad there's a deep valley past the end of the runway. This gave me an opportunity to push the yoke foreward to finally gain some respectable speed.

Density altitude! The Albuquerque airport is at 5,352 feet above sea level. At a temperature of 95 degrees F. the density altitude is a hefty 8,753 feet. The airplane "thinks" it's taking off from an airport at that elevation. Ever since then I've had a healthy respect for density altitude.

The mathematical formula used to determine density altitude is very long and complicated:

```
Density altitude = (145426*(1-(((288.15-pressure altitude
*.001981) /288.15) ^ 5.2563/((273.15+degrees centigrade)
/288.15)) ^ .235))
```

Once that formula is known, it's simple to arrive at the answer, based on known pressure altitude (the actual elevation) and the current temperature in degrees centigrade.

Once activated, the program asks for input:

```
Pressure altitude?
```

```
For temperature in centigrade press C
For temperature in Fahrenheit press F
Press C or F
```

```
Degrees Centigrade? or Degrees Fahrenheit?
```

Once these data have been entered, the program responds with:

```
Density altitude: 8753 feet
```

The program can also be used to determine the true airspeed at any flight level. It continues by asking:

```
Do you want to see the TAS at this altitude? (Y/N)
```

If you reply in the affirmative, it continues with:

```
Indicated airspeed? (knots)
```

Let's say it's 155 knots: the result will be displayed as:

```
True airspeed at 12,444 foot density altitude: 185 knots.
```

```
100 REM DENSITY ALTITUDE AND TAS
110 REM BASIC-80
120 GOTO 180
130 PRINT STRING$(80,45):RETURN
140 HOME:VTAB(10):RETURN
150 PRINT:PRINT:INPUT "Press >RETURN<  (Q to quit)  ",R$:IF R$="Q" THEN 160 ELSE RETURN
160 GOSUB 140:GOSUB 130:PRINT TAB(38)"End.":GOSUB 130:END
170 PRINT:INPUT "Entry correct?  (Y/N)  ",CORRECT$:RETURN
180 GOSUB 140:GOSUB 130:PRINT TAB(29)"DENSITY ALTITUDE AND TAS"
190 GOSUB 130:GOSUB 150
200 GOSUB 140:INPUT "Pressure altitude    ",PA
210 GOSUB 170:IF CORRECT$<>"N" THEN 220 ELSE GOSUB 140:GOTO 200
220 GOSUB 130
230 PRINT"For temperature in centigrade press  C":PRINT
240 PRINT"For temperature in Fahrenheit press  F":PRINT
250 INPUT "Press C or F    ",F$
260 IF F$="F" THEN 430
270 PRINT:INPUT "Degress centigrade  ",F:PRINT
280 GOSUB 170:IF CORRECT$<>"N" THEN 290 ELSE GOSUB 140:GOTO 270
290 C=288.15
300 D=(145426!X(1-(((C-PAX.001981)/C)^5.2563/(((273.15+F)/C))^.235))
310 D=CINT(D)
320 GOSUB 140:GOSUB 130
330 PRINT"Density altitude: ";D;" feet":GOSUB 130
340 INPUT "Do you want to see the TAS at this altitude? (Y/N)  ",YN$
350 IF YN$<>"Y" THEN 160 ELSE 360
360 GOSUB 140:INPUT "Indicated airspeed (knots)  ",IA:PRINT
370 GOSUB 170:IF CORRECT$<>"N" THEN 380 ELSE GOSUB 140:GOTO 360
380 Z=DX.00241
390 Y=IA+Z:Y=CINT(Y)
400 GOSUB 140:GOSUB 130
410 PRINT"True airspeed at ";D;" foot density altitude:  ";Y;" knots":GOSUB 130
420 GOSUB 150:GOTO 160
430 PRINT:INPUT "Degrees Fahrenheit    ",F:PRINT
440 GOSUB 170:IF CORRECT$<>"N" THEN 450 ELSE GOSUB 140:GOTO 430
450 F=(F-32)/1.8
460 GOTO 290
```

Figure 39.1 The density altitude and true air speed program.

The program (Figure 39.1) is quite short. Lines 100–250 are used as input lines, and lines 430–460 are used if the temperature is to be input in degrees Fahrenheit, converting that figure to degrees Centigrade. Lines 290–310 perform the calculation for density altitude, and lines 340–420 perform the true airspeed calculation and display the result.

Converting Local Time to Greenwich Mean Time

Flight plans, especially IFR flight plans, are supposed to use Greenwich Mean Time (Zulu Time) instead of local standard or daylight time. This avoids the confusion that can result when we're dealing with the different North American time zones. While conversion presents no serious problem, it's vexing when we're trying to remember whether to add or deduct an hour to allow for daylight saving time.

The program (Figure 40.1) performs the conversions to or from Zulu Time relative to any of the four major standard, as well as daylight, time zones. When activated, it asks:

Zulu Time Conversion

For INPUT use 24-hour clock!

----

1   Convert TO Zulu time
2   Convert FROM Zulu time

Which?

Time zone to be converted:

----

1   Eastern Standard Time
2   Central Standard Time
3   Mountain Standard Time
4   Pacific Standard Time

```
5  Eastern Daylight Time
6  Central Daylight Time
7  Mountain Daylight Time
8  Pacific Daylight Time
```

Which?

Depending on your choice, the computer then asks that the time be keyed in:

```
Local time? (hrs.min)  or  Zulu time? (hrs.min)
```

separating the hours and minutes by a decimal point rather than a colon. Once the time is entered, the program displays the converted time, identifying it as Zulu or local time.

```
100 REM ZULU TIME CONVERSION
110 REM BASIC-80
120 GOTO 200
130 PRINT STRING$(80,45):RETURN
140 HOME:VTAB(10):RETURN
150 PRINT:PRINT:INPUT "Press >RETURN<  (Q to quit)  ",Y$:IF Y$="Q" THEN 160 ELSE RETURN
160 GOSUB 140:GOSUB 130:PRINT TAB(38)"End.":GOSUB 130:END
170 PRINT:INPUT "Another conversion? (Y/N)  ",AC$:IF AC$="Y" THEN 230 ELSE 160
180 GOSUB 140:INPUT "Local time? (hrs.min)   ",LT:GOSUB 130:RETURN
190 GOSUB 140:INPUT "Zulu time? (hrs.min)    ",ZT:GOSUB 130:RETURN
200 GOSUB 140:GOSUB 130
210 PRINT TAB(25)"Zulu Time Conversion":GOSUB 130
220 PRINT"For INPUT use 24-hour clock!":GOSUB 130:GOSUB 150
230 GOSUB 140
240 PRINT 1,"Convert TO   Zulu time"
250 PRINT 2,"Convert FROM Zulu time":GOSUB 130
260 INPUT "Which?  ",A:GOSUB 140
270 PRINT"Time zone to be converted:":GOSUB 130
280 PRINT 1,"Eastern  Standard Time"
290 PRINT 2,"Central  Standard Time"
300 PRINT 3,"Mountain Standard Time"
310 PRINT 4,"Pacific  Standard Time":GOSUB 130
320 PRINT 5,"Eastern  Daylight Time"
330 PRINT 6,"Central  Daylight Time"
340 PRINT 7,"Mountain Daylight Time"
350 PRINT 8,"Pacific  Daylight Time":GOSUB 130
360 INPUT "Which?   ",N:PRINT
370 ON N GOTO 410,570,580,590,600,380,390,400
380 Q=7:GOTO 420
390 Q=8:GOTO 420
400 Q=9:GOTO 420
410 Q=5
420 IF Q=2 THEN 500
430 GOSUB 180
440 ZT=LT+Q
450 TZ=INT(ZT):ZT=ZT-TZ
460 IF TZ>24 THEN TZ=TZ-24
470 TT=TZ+ZT
```

continued

```
480 PRINT USING "Zulu time:              ##.##";TT
490 GOSUB 130:GOTO 170
500 GOSUB 190
510 LT=ZT-Q
520 TL=INT(LT):LT=LT-TL
530 IF TL<1 THEN TL=TL+24
540 TT=TL+LT
550 PRINT USING "Local time:             ##.##";TT
560 GOSUB 130:GOTO 170
570 Q=6:GOTO 420
580 Q=7:GOTO 420
590 Q=8:GOTO 420
600 Q=6:GOTO 420
```

Figure 40.1 The program that converts local time to Greenwich Mean Time and vice versa.

Lines 100–190 represent a number of subroutines used throughout the program. Lines 200–370 present the choices, sending the computer to the appropriate line numbers. Lines 380–600 perform the conversions, with lines 480 and 550 displaying the results.

CHAPTER FORTY-ONE

Determining the Advantages of Aircraft Ownership

Operating an aircraft is not cheap. Whether aircraft ownership is justified depends largely on your travel needs. This program (Figure 41.1) is divided into four subprograms and deals with various financial aspects of aircraft ownership. When your average annual travel needs are keyed in, it displays recommendations about the type of aircraft, if any, that might be best suited for your purposes. A subprogram computes the costs associated with aircraft ownership: another subprogram assigns dollar values to benefits of airplane ownership; and the last subprogram may be used to determine the value of executive time.

```
100 REM BUSINESS AIRCRAFT COST ANALYSIS
110 REM BASIC-80
120 GOTO 180
130 PRINT STRING$(80,45):RETURN
140 HOME:VTAB(10):RETURN
150 PRINT:INPUT "Press >RETURN< (Q to quit) ",R$:IF R$="Q" THEN 160 ELSE RETURN
160 GOSUB 140:GOSUB 130:PRINT TAB(38)"End.":GOSUB 130:END
170 PRINT:INPUT "Entries correct?  (Y/N)  ",CORRECT$:RETURN
180 TS$="Time saved by using business aircraft:    "
190 RE$="Reliability of available transportation:  "
200 CO$="Control over facilities away from home:   "
210 PR$="Privacy while traveling:                  "
220 ES$="Prestige attached to aircraft ownership   "
230 FL$="Scheduling flexibility:                   "
240 SA$="Safety, including terrorism protection:   "
250 MF$="Miscellaneous fringe benefits:            "
260 MA$="Marketing advantages:                     "
270 PE$="Ability to attract top-level personnel:   "
280 GOSUB 140:PRINT TAB(25)"BUSINESS AIRCRAFT ANALYSES":GOSUB 130
290 PRINT TAB(15)"Select one of four programs.":GOSUB 130
300 PRINT 1,"Do company travel needs justify aircraft acquisition?"
310 PRINT 2,"Determining the cost of business aircraft operation."
320 PRINT 3,"Determining the benefits derived from aircraft ownership."
330 PRINT 4,"Determining the value of executive time.":GOSUB 130
340 PRINT 5,"Exit program":GOSUB 130
350 INPUT "Which?  ",A:GOSUB 140
360 ON A GOTO 520,920,1060,370,160
370 GOSUB 140:GOSUB 130
```

continued

```
380 PRINT TAB(7)"DETERMINING THE VALUE OF ONE HOUR OF AN EXECUTIVE'S OR":PRINT
390 PRINT TAB(7)"AN EMPLOYEE'S TIME TO THE EMPLOYER.":GOSUB 130:GOSUB 150:GOSUB 140
400 INPUT "Annual wage (including fringe benefits)   $",AW
410 HW=AW/2000:P=HW/2:V=HW*P:GOSUB 130
420 PRINT USING "Actual wage per hour:                    $$###.##";HW
430 GOSUB 130:AV=V*2000:MV=AV/12:WV=AV/52:DV=AV/250
440 PRINT USING "Value per year to employer:          $$########,.##";AV
450 PRINT USING "Value per month to employer:          $$#######,.##";MV
460 PRINT USING "Value per week to empoyer:            $$######,.##";WV
470 PRINT USING "Value per work day to employer:        $$#####,.##";DV
480 PRINT USING "Value per hour to employer:           $$#####,.##";V
490 GOSUB 130:GOSUB 150
500 GOSUB 140:INPUT "Do you want another evaluation? (Y/N) ",Y$
510 IF Y$="Y" THEN 370 ELSE 280
520 GOSUB 140
530 PRINT TAB(20)"Key in company travel needs per year.":GOSUB 130
540 INPUT "Average trip distance (miles)?            ",ZN
550 INPUT "How many such trips per year?             ",ZA
560 INPUT "Longest trips made occasionally (miles)?  ",ZL
570 INPUT "How many such trips?                      ",ZB
580 INPUT "Average number of passengers?             ",ZP
590 INPUT "IFR or night flying needed? (Y/N)         ",ZI$
600 GOSUB 170:IF CORRECT$<>"N" THEN 610 ELSE GOSUB 140:GOTO 540
610 GOSUB 140
620 IF ZI$="Y" THEN 640
630 PRINT TAB(10)"Single-engine aircraft acceptable.":GOTO 650
640 PRINT TAB(10)"Multi-engine aircraft suggested.":PRINT
650 IF ZN>=500 THEN 710
660 IF ZA>=40 THEN 730
670 IF ZP>=4 THEN 750
680 IF ZB>=10 THEN 770
690 IF ZN<200 AND ZA<20 THEN 800
700 GOTO 790
710 PRINT TAB(10)"Pressurization is suggested.":GOSUB 130:GOSUB 150
720 GOTO 660
730 PRINT:PRINT TAB(10)"You may need two aircraft":GOSUB 130:GOSUB 150
740 GOTO 670
750 PRINT TAB(10)"Cabin-size twin aircraft suggested":GOSUB 130:GOSUB 150
760 GOTO 680
770 PRINT TAB(10)"Minimum range with IFR reserves:   ";ZL:GOSUB 130:GOSUB 150
780 PRINT TAB(10)"Turbine aircraft suggested.":GOSUB 130:GOSUB 150
790 GOTO 810
800 GOSUB 130:PRINT TAB(10)"Aircraft ownership not advisable.":GOSUB 150
810 GOSUB 140:PRINT TAB(10)"To determine average trip times for different types":PRINT
820 PRINT TAB(10)"of aircraft, key in the cruise speeds.":PRINT
830 INPUT "          Cruise speed? (knots or mph)    ",KT
840 HM=ZN/KT:HS=ZL/KT:HY=ZN*ZA:J=(ZL*ZB)-(ZN*ZB):J=J+HY:HY=J/KT:HY=HY*2:GOSUB 130
850 HM=INT(HM*10+.5)/10:HX=INT(HM):XH=(HM-HX)*.6:XH=XH*100
860 PRINT TAB(10)"Average trip time:      ";HX;" hours and ";XH;" minutes.":PRINT
870 HS=INT(HS*10+.5)/10:HX=INT(HS):XH=(HS-HX)*.6:HS=HX+XH:XH=XH*100
880 PRINT TAB(10)"Time for long trips:     ";HX;" hours and ";XH;" minutes.":PRINT:HY=INT(HY)
890 PRINT TAB(10)"Hours per year based on entered data:   ";HY;" hours":GOSUB 130
900 INPUT "          Examine other cruise speeds? (Y/N)   ",CS$
910 IF CS$="Y" THEN GOSUB 140:GOTO 830 ELSE 280
920 GOSUB 140:Q=2:GOTO 1430
930 GOSUB 140
940 HTAB(10):PRINT USING "Total aircraft miles:    ########,.#";NM
950 HTAB(10):PRINT USING "Total passenger miles:   ########,.#";PM
960 HTAB(10):PRINT USING "Pre-tax cost per hour:    $$#####,.##";TC
970 HTAB(10):PRINT USING "Pre-tax cost per year:   $$######,.##";GC
980 HTAB(10):PRINT USING "Cost/year after taxes:   $$######,.##";NC
990 HTAB(10):PRINT USING "Cost/hour after taxes:    $$#####,.##";AH
1000 HTAB(10):PRINT USING "Cost per aircraft mile:  $$#####,.##";AM
1010 HTAB(10):PRINT USING "Cost per passenger mile: $$#####,.##";PN
1020 HTAB(10):PRINT USING "Dollar value/hrs. saved: $$#####,.##";HR
```

```
1030 HTAB(10):PRINT USING "Cost/authorized user:    $$######,.##";CU:GOSUB 130
1040 INPUT "Do you want to repeat (Y/N)    ",Y$
1050 IF Y$="Y" THEN 920 ELSE 280
1060 GOSUB 140:TS=5:RE=5:CO=5:PR=5:ES=5:FL=5:SA=5:MF=5:MA=5:PE=5:Q=3
1070 PRINT TAB(10)"Each category of benefits is given a percentage value":PRINT
1080 PRINT TAB(10)"based on average assessment.  You may change those":PRINT
1090 PRINT TAB(10)"values by typing in your own assessment.  Otherwise,":PRINT
1100 PRINT TAB(10)"type >RETURN< to accept displayed value and go on.":GOSUB 130
1110 PRINT TS$;"  20%";
1120 INPUT TS:IF TS<>0 THEN 1140
1130 TS=20
1140 GOSUB 130:PRINT RE$;"  18%";
1150 INPUT RE:IF RE<>0 THEN 1170
1160 RE=18
1170 GOSUB 130:PRINT CO$;"  16%";
1180 INPUT CO:IF CO<>0 THEN 1200
1190 CO=16
1200 GOSUB 130:PRINT PR$;"  13%";
1210 INPUT PR:IF PR<>0 THEN 1230
1220 PR=13
1230 GOSUB 130:PRINT ES$;"  10%";
1240 INPUT ES:IF ES<>0 THEN 1260
1250 ES=10
1260 GOSUB 130:PRINT FL$;"   9%";
1270 INPUT FL:IF FL<>0 THEN 1290
1280 FL=9
1290 GOSUB 130:PRINT SA$;"   7%";
1300 INPUT SA:IF SA<>0 THEN 1320
1310 SA=7
1320 GOSUB 130:PRINT MF$;"   4%";
1330 INPUT MF:IF MF<>0 THEN 1350
1340 MF=4
1350 GOSUB 130:PRINT MA$;"   2%";
1360 INPUT MA:IF MA<>0 THEN 1380
1370 MA=2
1380 GOSUB 130:PRINT PE$;"   1%";
1390 INPUT PE:IF PE<>0 THEN 1410
1400 PE=1
1410 GOSUB 130
1420 GOSUB 140
1430 INPUT "Number of users?                       ",US
1440 INPUT "Aircraft hours per year?               ",HY
1450 INPUT "Average cruise speed?                  ",CS
1460 INPUT "Average number. of passengers?         ",PX
1470 INPUT "Number of aircraft trips?              ",TR
1480 INPUT "No. of trips by co. personnel?         ",PT
1490 INPUT "No. of trips by customers?             ",CT
1500 INPUT "Estim. number man hours saved?         ",MH
1510 INPUT "Estim. $=value per man hour?         $",VH
1520 INPUT "$=value of airlines not used?        $",AS
1530 INPUT "$=value of R.O.Ns saved?             $",RS
1540 INPUT "Company tax bracket?                 %",TB
1550 INPUT "Fixed costs/aircraft hour?           $",FC
1560 INPUT "Variable costs/aircraft hour?        $",VC
1570 GOSUB 170:IF CORRECT$<>"N" THEN 1580 ELSE GOSUB 140:GOTO 1430
1580 GOTO 1630
1590 TS=(NC/MH)*(TS/100):RE=(NC/(PT+CT))*(RE/100):CO=(NC/TR)*(CO/100)
1600 PR=((NC/HY)/PX)*(PR/100):ES=NC*(ES/100):FL=(NC/(PT+CT))*(FL/100)
1610 SA=(NC/TR)*(SA/100):MF=(NC/US)*(MF/100):MA=(NC/CT)*(MA/100)
1620 PE=(NC/TR)*(PE/100):GOTO 1670
1630 NM=HY*CS:PM=PX*(HY*CS):TC=FC+VC:SS=(VH*MH)+AS+RS:GC=(TC*HY)-AS-RS
1640 NC=GC-(GC*(TB/100)):AH=NC/HY:AM=NC/NM:PN=NC/PM:HR=VH*MH:CU=NC/US
```

continued

```
1650 IF Q=3 THEN 1590
1660 IF Q=2 THEN 930
1670 GOSUB 140
1680 TS=INT(TS*100+.5)/100:RE=INT(RE*100+.5)/100:CO=INT(CO*100+.5)/100
1690 PR=INT(PR*100+.5)/100:ES=INT(ES*100+.5)/100:FL=INT(FL*100+.5)/100
1700 SA=INT(SA*100+.5)/100:MF=INT(MF*100+.5)/100:MA=INT(MA*100+.5)/100
1710 PE=INT(PE*100+.5)/100:GOSUB 140
1720 PRINT TS$;
1730 PRINT USING "$$#####,.##";TS;
1740 PRINT TAB(60)" per hour saved"
1750 PRINT RE$;
1760 PRINT USING "$$#####,.##";RE;
1770 PRINT TAB(60)" per passenger trip"
1780 PRINT CO$;
1790 PRINT USING "$$#####,.##";CO;
1800 PRINT TAB(60)" per aircraft trip"
1810 PRINT PR$;
1820 PRINT USING "$$#####,.##";PR;
1830 PRINT TAB(60)" per flight hour"
1840 PRINT ES$;
1850 PRINT USING "$$#####,.##";ES;
1860 PRINT TAB(60)" per year"
1870 PRINT FL$;
1880 PRINT USING "$$#####,.##";FL;
1890 PRINT TAB(60)" per passenger trip"
1900 PRINT SA$;
1910 PRINT USING "$$#####,.##";SA;
1920 PRINT TAB(60)" per aircraft trip"
1930 PRINT MF$;
1940 PRINT USING "$$#####,.##";MF;
1950 PRINT TAB(60)" per user"
1960 PRINT MA$;
1970 PRINT USING "$$#####,.##";MA;
1980 PRINT TAB(60)" per customer"
1990 PRINT PE$;
2000 PRINT USING "$$#####,.##";PE;
2010 PRINT TAB(60)" per landing":GOSUB 150
2020 GOSUB 140:INPUT "Do you wish to repeat? (Y/N)   ";Z$
2030 IF Z$="Y" THEN 1060:ELSE 280
```

Figure 41.1 The program determines whether aircraft ownership is justified, computes aircraft operating costs, displays benefits derived from aircraft ownership, and computes the value of executive or employee time.

Starting the program produces the main menu. Select any one of the subprograms:

```
                    BUSINESS AIRCRAFT ANALYSES
-------------------------------------------------------------------------------
Select one of four programs:
-------------------------------------------------------------------------------

    1   Do company travel needs justify aircraft acquisition?
    2   Determining the cost of business aircraft operation.
    3   Determining the benefits derived from aircraft ownership.
    4   Determining the value of executive time.

-------------------------------------------------------------------------------
```

```
 5 Exit program
-------------------------------------------------------------------------------
```
Which?

If you select the first choice, the program prompts for inputs:

```
Key in company travel needs per year:
-------------------------------------------------------------------------------
Average trip distance (miles)?
How many such trips per year?
Longest trips made occasionally (miles)?
How many such trips?
Average number of passengers?
IFR or night flying needed? (Y/N)
```

The program then uses the answers to these questions to display one or several of the following recommendations:

```
Single-engine aircraft acceptable.
Multi-engine aircraft suggested.
Pressurization suggested.
You may need two aircraft.
Cabin-size twin aircraft suggested.
Minimum range with IFR reserves:
Turbine aircraft suggested.
Aircraft ownership not advisable.
```

At this point you have the option of determining the trip duration for different types of aircraft:

```
To determine average trip duration for different types of air-
craft, key in the cruise speeds.
-------------------------------------------------------------------------------
    Cruise speed? (knots or mph)
-------------------------------------------------------------------------------
    Average trip time:          2 hours and 14 minutes.
    Time for long trips:        5 hours and 27 minutes.
    Hours per year based on entered data: 413
-------------------------------------------------------------------------------
    Examine other cruise speeds? (Y/N)
```

The second choice from the main menu computes the annual cost of operating an aircraft, the data for which must be keyed in based on a series of prompts:

```
Number of users?
Aircraft hours per year?
Average cruise speed?
Average number of passengers?
Number of aircraft trips?
No. of trips by co. personnel?
No. of trips by customers?
Estim. number man hours saved?
Estim. $=value per man hour?            $
$=value of airlines not used?          $
$=value of R.O.Ns saved?               $
Company tax bracket$                   %
Fixed costs/aircraft hour?             $
Variable costs/aircraft hour?          $
```

Once all these estimates have been entered, the computer performs the necessary calculations and produces the results in this format:

```
Total aircraft miles:               110,000
Total passenger miles:              330,000
Pre-tax cost per hour:               $75.00
Pre-tax cost per year:           $26,250.00
Cost/year after taxes:           $17,062.50
Cost/hour after taxes:               $48.75
Cost per aircraft mile:               $0.16
Cost per passenger mile:              $0.05
Dollar value/hrs. saved:         $24,500.00
Cost/authorized user:             $3,412.50
```

The figures used here are based on an annual use of 350 hours for a turbine aircraft with an average cruise speed of 315 knots, operated by a company in the 35-percent tax bracket (though the $75 hourly cost is admittedly overly-optimistic).

The third selection from the menu displays ten categories of benefits:

```
Time saved by using business aircraft
Reliability of available transportation
Control over facilities away from home
Privacy while traveling
Prestige attached to aircraft ownership
Scheduling flexibility
```

```
Safety, including terrorism protection
Miscellaneous fringe benefits
Marketing advantages
Ability to attract top-level personnel
```

All these subjects are given a percentage-of-importance rating which can be changed by the user. Once these percentage ratings are accepted or adjusted, the computer asks for the same inputs shown above, after which it once more displays the list of benefits adding a dollar figure to each.

The last subprogram determines the estimated value of time, based on the annual remuneration of an executive or employee. The formula was worked out many years ago by a major management consultant firm. The basic idea is that any person who is being paid by a company must return X times his or her salary in terms of contribution to the company in order to permit the company to cover overhead and myriads of other costs and end up with a profit.

The subprogram asks that you key in annual remuneration including fringe benefits, and it then displays the value per year, month, week, work day and hour of the executive or employee to the company. That value can be used in the other subprograms, where the value of hours saved is included in the determination of the final cost of aircraft ownership.

CHAPTER FORTY-TWO

Aircraft Expense Record Program

Designed to keep a permanent record of aircraft-related expenses, this program differentiates between the costs of individual trips, and maintenance and repair costs. Expense data are stored in two random-access files and can be recalled at any time. In addition to these two data files, the program creates two sequential files used internally to keep track of the data entries, numbering them consecutively from 1 to 999.

When the program is activated, it displays a series of menus:

```
                              Menu:
-----------------------------------------------------------------
  1   Enter new expense data
  2   Review previous expense data
-----------------------------------------------------------------
  3   Exit program
-----------------------------------------------------------------
Which?
```

If the first choice is selected, it continues with:

```
Enter expense data for:
-----------------------------------------------------------------
  1   Individual trips
  2   Maintenance and non-trip-related expenses
-----------------------------------------------------------------
  3   Exit program
-----------------------------------------------------------------
Which?
```

If the data to be entered are related to a given trip, the program displays a series of prompts:

```
Flight description?    (Example: JFK-MSY)
Starting date of trip?    (Example: 9/30/85)
Distance in nautical miles?
Fuel consumed?    (Gallons or pounds)
Average fuel cost per unit?                                    $
Cost of other consumables?                                     $
Landing fees?                                                  $
Cost of tiedown, hangar?                                       $
Misc. expenses?                                                $
```

When these data have been keyed in, the program computes the total fuel cost for the trip and the total cost of the entire trip, displaying these totals (here using arbitrary numbers):

```
Total fuel cost:                                          $120.00
Total entire trip:                                        $215.00
```

These data are stored in one of the data files, and can be examined or printed during any subsequent run of the program.

If maintenance and non-trip-related expenses are to be entered, another series of prompts is displayed:

```
Date?    (Example 2/1/85 or 10/15/85)
Airframe maintenance expense?                                  $
Engine maintenance expense?                                    $
Avionics maintenance expense?                                  $
Other expense?                                                 $
```

The entries are sent to a special data file for subsequent recall.

If the second choice, "Review previous expense data" is selected from the main menu, the next menu looks like this:

```
                    Review expense data for:
-----------------------------------------------------------------

    1  Flights
    2  Maintenance etc.

=================================================================

    3  Exit program
-----------------------------------------------------------------

Which?
```

If the first choice is selected, the display responds with one more menu:

Review expense data for:

---

1 A trip by date
2 A trip by flight description
3 All trips to date

---

4 Review all flight descriptions and dates

---

5 Exit program

Which?

When the first or second choice is called up, the program asks that either the date or the flight description for the flight in question be keyed in:

Starting date of flight? (Example: 6/5/85 or 11/12/85)

or

Flight description? (Example: JFK–LAX)

The computer then searches for the group of flight-related expenses that includes either the date or the flight description, displaying these data or sending them to the line printer. (See Figure 42.1).

```
Flight description:                   SAF-LAX
Starting date:                        2/18/85
Distance in nautical miles:           800
Fuel consumed:  (Gallons or pounds)   60
Average fuel cost per unit:             $2.00
Cost of other consumables:            $15.00
Landing fees:                         $10.00
Cost of tiedown, hangar:              $20.00
Misc. expenses:                       $50.00
-------------------------------------------------------------
Total fuel cost:                     $120.00
Total entire trip:                   $215.00
```

Figure 42.1 A sample print-out showing the expense data for a single flight.

If the computer was asked to display and/or print the previously-entered expense data for all trips, it simply starts at the first entry and displays the data sequentially, each time asking whether the next group of data is to be displayed or printed.

Since it may be difficult to remember the dates for specific flights, the fourth choice from the menu causes the dates and related flight descriptions for all flights to be displayed or printed (see Figure 42.2).

If previously-entered maintenance costs are to be reviewed, the program asks no further questions. It simply displays all previous entries along with the consecutive totals, as shown in Figure 42.3.

```
Flight description:                   SAF-LAX
Starting date:                        2/18/85
-------------------------------------------------------------
Flight description:                   LAX-SFO
Starting date:                        2/20/85
-------------------------------------------------------------
```

Figure 42.2 A sample print-out showing all flights for which data were recorded and saved.

```
Date:                                     2/18/85
Airframe maintenance expense:               $739.00
Engine maintenance expense:                 $955.00
Avionics maintenance expense:               $222.00
Other expense:                              $100.00
-----------------------------------------------------------------
Total Airframe expense to this point:       $739.00
Total Engine expense to this point:         $955.00
Total Avionics expense to this point:       $222.00
Total other expense to this point:          $100.00
-----------------------------------------------------------------
Total to this point:                      $2,016.00
-----------------------------------------------------------------
Date:                                     2/25/85
Airframe maintenance expense:               $500.00
Engine maintenance expense:                 $300.00
Avionics maintenance expense:               $250.00
Other expense:                              $175.00
-----------------------------------------------------------------
Total Airframe expense to this point:     $1,239.00
Total Engine expense to this point:       $1,255.00
Total Avionics expense to this point:       $472.00
Total other expense to this point:          $275.00
-----------------------------------------------------------------
Total to this point:                      $3,241.00
-----------------------------------------------------------------
```

Figure 42.3 A sample print-out of the maintenance and repair cost records produced by the program.

```
100 REM AIRCRAFT EXPENSE FILE PROGRAM  FILENAME.SUB/F:4
110 REM BASIC-80
120 GOTO 200
130 PRINT STRING$(80,45):RETURN
140 LPRINT STRING$(60,45):RETURN
150 HOME:VTAB(10):RETURN
160 PRINT:INPUT "Press >RETURN<  (Q to quit)  ",R$
170 IF R$="Q" THEN 180 ELSE RETURN
180 GOSUB 150:GOSUB 130:PRINT TAB(38)"End.":GOSUB 130:END
190 PRINT:INPUT "Entries correct?  (Y/N)  ",CORRECT$:RETURN
200 GOSUB 150
210 PRINT TAB(20)"AIRCRAFT EXPENSE RECORDS PROGRAM"
220 PRINT TAB(20)STRING$(32,42):PRINT
230 PRINT TAB(10)"Menu:":GOSUB 130
240 PRINT 1,"Enter new expense data"
250 PRINT 2,"Review previous expense data":GOSUB 130
260 PRINT 3,"Exit program":GOSUB 130
270 INPUT "Which?   ",WHICH:GOSUB 150
280 ON WHICH GOTO 290,740,180
290 PRINT TAB(10)"Enter expense data for:":GOSUB 130
300 PRINT 1,"Individual trips"
```

```
310 PRINT 2,"Maintenance and non-trip-related expenses":GOSUB 130
320 PRINT 3,"Exit program":GOSUB 130
330 INPUT "Which?  ",WH:GOSUB 150
340 ON WH GOTO 350,600,180
350 ON ERROR GOTO 370
360 OPEN "I",#1,"RECORD.NUM":INPUT #1,RN%:CLOSE #1
370 RN%=RN%+1:OPEN "O",#1,"RECORD.NUM":PRINT #1,RN%:CLOSE #1
380 OPEN "R",#2,"TRIP.DTA",79
390 FIELD #2, 3 AS R$,7 AS F$,8 AS Z$,4 AS D$,5 AS L$,5 AS A$,7 AS C$,6 AS G$,6 AS H$,7
AS M$,10 AS Y$,10 AS X$
400 INPUT "Flight description? (Example: JFK-MSY)     ",FLIGHT$
410 INPUT "Starting date of trip (Example: 9/30/85)   ",DATE$
420 INPUT "Distance in nautical miles?                 ",DIST$
430 INPUT "Fuel consumed?  (Gallons or pounds)        ",FUEL$
440 INPUT "Average fuel cost per unit?               $",COSTF$
450 INPUT "Cost of other consumables?                $",COSTC$
460 INPUT "Landing fees?                             $",LDG$
470 INPUT "Cost of tiedown, hangar?                  $",HANGAR$
480 INPUT "Misc. expenses?                           $",MISC$:GOSUB 130
490 GOSUB 190:IF CORRECT$<>"N" THEN 500 ELSE GOSUB 150:GOTO 400
500 FUEL=VAL(FUEL$):COSTF=VAL(COSTF$):COSTC=VAL(COSTC$):LDG=VAL(LDG$)
510 HANGAR=VAL(HANGAR$):MISC=VAL(MISC$)
520 TOTFUEL=FUEL*COSTF:TOTOTHER=COSTC+LDG+HANGAR+MISC:TOTAL=TOTFUEL+TOTOTHER
530 GOSUB 130:PRINT USING "Total fuel cost:       $$######.##";TOTFUEL
540 PRINT USING "Total entire trip:     $$######.##";TOTAL
550 LSET R$=MKI$(RN%):LSET F$=FLIGHT$:LSET Z$=DATE$:LSET D$=DIST$
560 LSET L$=FUEL$:LSET A$=COSTF$
570 LSET C$=COSTC$:LSET G$=LDG$:LSET H$=HANGAR$:LSET M$=MISC$
580 LSET Y$=MKI$(TOTFUEL):LSET X$=MKI$(TOTAL)
590 PUT #2,RN%:CLOSE #2:GOSUB 160:GOSUB 150:GOTO 230
600 ON ERROR GOTO 620
610 OPEN "I",#3,"MAINT.NUM":INPUT #3,MN%:CLOSE #3
620 MN%=MN%+1:OPEN "O",#3,"MAINT.NUM":PRINT #3,MN%:CLOSE #3
630 OPEN "R",#4,"MAINT.DTA",44
640 FIELD #4,3 AS N$,8 AS D$,8 AS F$,8 AS E$,8 AS A$,8 AS O$
650 INPUT "Date?        (Example 2/1/85 or 10/15/85)  ",DAT$
660 INPUT "Airframe maintenance expense?            $",FRAME$
670 INPUT "Engine maintenance expense?              $",ENGIN$
680 INPUT "Avionics maintenance expense?            $",AVION$
690 INPUT "Other expense?                           $",OTHER$
700 GOSUB 190:IF CORRECT$<>"N" THEN 710 ELSE GOSUB 150:GOTO 650
710 LSET N$=MKI$(MN%):LSET D$=DAT$:LSET F$=FRAME$:LSET E$=ENGIN$
720 LSET A$=AVION$:LSET O$=OTHER$
730 PUT #4,MN%:CLOSE #4:GOSUB 130:GOSUB 150:GOTO 230
740 PRINT TAB(10)"Review expense data for:":GOSUB 130
750 PRINT 1,"Flights"
760 PRINT 2,"Maintenance, etc.":GOSUB 130
770 PRINT 3,"Exit program":GOSUB 130
780 INPUT "Which?    ",WHCH:GOSUB 150
790 ON WHCH GOTO 800,1970,180
800 PRINT TAB(10)"Review expense data for:":GOSUB 130
810 PRINT 1,"A trip by date"
820 PRINT 2,"A trip by flight description"
830 PRINT 3,"All trips to date":GOSUB 130
840 PRINT 4,"Review all flight descriptions and dates":GOSUB 130
850 PRINT 5,"Exit program":GOSUB 130
860 INPUT "Which?  ",WCH:GOSUB 150
870 ON WCH GOTO 880,1220,1540,1860,180
880 INPUT "Starting date of flight? (Example:6/5/85 or 11/12/85)   ",START$
890 OPEN "I",#1,"RECORD.NUM":INPUT #1,RN%:CLOSE #1
900 OPEN "R",#2,"TRIP.DTA",79
910 FIELD #2, 3 AS R$,7 AS F$,8 AS Z$,4 AS D$,5 AS L$,5 AS A$,7 AS C$,6 AS G$,6 AS H$,7
AS M$,10 AS Y$,10 AS X$
920 FOR X=1 TO RN%:GET #2,X
```

```
930 L=LEN(START$)
940 IF START$<>LEFT$(Z$,L) THEN 1090 ELSE 950
950 A=VAL(A$):C=VAL(C$):G=VAL(G$):H=VAL(H$):M=VAL(M$)
960 PRINT:INPUT "Printout? (Y/N)  ",PR$:PRINT:IF PR$<>"Y" THEN 970 ELSE 1100
970 PRINT"Flight description:                        ";F$
980 PRINT"Starting date:                             ";Z$
990 PRINT"Distance in nautical miles:                ";D$
1000 PRINT"Fuel consumed: (Gallons or pounds)        ";L$
1010 PRINT USING "Average fuel cost per unit:                $$##.##";A
1020 PRINT USING "Cost of other consumables:                 $$####,.##";C
1030 PRINT USING "Landing fees:                              $$###.##";G
1040 PRINT USING "Cost of tiedown, hangar:                   $$###,.##";H
1050 PRINT USING "Misc. expenses:                            $$####,.##";M:GOSUB 130
1060 PRINT USING "Total fuel cost:                           $$#####,.##";CVI(Y$)
1070 PRINT USING "Total entire trip:                         $$#####,.##";CVI(X$)
1080 GOSUB 130:GOSUB 160:GOSUB 150:GOTO 230
1090 NEXT X:CLOSE #2:GOSUB 130:GOSUB 160:GOSUB 150:GOTO 230
1100 LPRINT"Flight description:                       ";F$
1110 LPRINT"Starting date:                            ";Z$
1120 LPRINT"Distance in nautical miles:               ";D$
1130 LPRINT"Fuel consumed: (Gallons or pounds)        ";L$
1140 LPRINT USING "Average fuel cost per unit:               $$##.##";A
1150 LPRINT USING "Cost of other consumables:                $$####,.##";C
1160 LPRINT USING "Landing fees:                             $$###.##";G
1170 LPRINT USING "Cost of tiedown, hangar:                  $$###,.##";H
1180 LPRINT USING "Misc. expenses:                           $$####,.##";M:GOSUB 140
1190 LPRINT USING "Total fuel cost:                          $$#####,.##";CVI(Y$)
1200 LPRINT USING "Total entire trip:                        $$#####,.##";CVI(X$)
1210 GOTO 1080
1220 INPUT "Flight description? (Example: JFK-LAX)  ",DESCR$
1230 OPEN "I",#1,"RECORD.NUM":INPUT #1,RN%:CLOSE #1
1240 OPEN "R",#2,"TRIP.DTA",79
1250 FIELD #2, 3 AS R$,7 AS F$,8 AS Z$,4 AS D$,5 AS L$,5 AS A$,7 AS C$,6 AS G$,6 AS H$,7
AS M$,10 AS Y$,10 AS X$
1260 FOR X=1 TO RN%:GET #2,X
1270 IF DESCR$<>F$ THEN 1550 ELSE 1280
1280 A=VAL(A$):C=VAL(C$):G=VAL(G$):H=VAL(H$):M=VAL(M$)
1290 PRINT:INPUT "Printout? (Y/N)   ",PR$:PRINT:IF PR$<>"Y" THEN 970 ELSE 1100
1300 PRINT"Flight description:                       ";F$
1310 PRINT"Starting date:                            ";Z$
1320 PRINT"Distance in nautical miles:               ";D$
1330 PRINT"Fuel consumed: (Gallons or pounds)        ";L$
1340 PRINT USING "Average fuel cost per unit:               $$##.##";A
1350 PRINT USING "Cost of other consumables:                $$####,.##";C
1360 PRINT USING "Landing fees:                             $$###.##";G
1370 PRINT USING "Cost of tiedown, hangar:                  $$###,.##";H
1380 PRINT USING "Misc. expenses:                           $$####,.##";M:GOSUB 130
1390 PRINT USING "Total fuel cost:                          $$#####,.##";CVI(Y$)
1400 PRINT USING "Total entire trip:                        $$#####,.##";CVI(X$)
1410 GOTO 1080
1420 NEXT X:CLOSE #2:GOSUB 130:GOSUB 160:GOSUB 150:GOTO 230
1430 LPRINT"Flight description:                       ";F$
1440 LPRINT"Starting date:                            ";Z$
1450 LPRINT"Distance in nautical miles:               ";D$
1460 LPRINT"Fuel consumed: (Gallons or pounds)        ";L$
1470 LPRINT USING "Average fuel cost per unit:               $$##.##";A
1480 LPRINT USING "Cost of other consumables:                $$####,.##";C
1490 LPRINT USING "Landing fees:                             $$###.##";G
1500 LPRINT USING "Cost of tiedown, hangar:                  $$###,.##";H
1510 LPRINT USING "Misc. expenses:                           $$####,.##";M:GOSUB 140
1520 LPRINT USING "Total fuel cost:                          $$#####,.##";CVI(Y$)
1530 LPRINT USING "Total entire trip:                        $$#####,.##";CVI(X$)
1540 GOTO 1080
1550 NEXT X:CLOSE #2:GOSUB 150:GOTO 230
```

```
1560 OPEN "I",#1,"RECORD.NUM":INPUT #1,RN%:CLOSE #1
1570 OPEN "R",#2,"TRIP.DTA",79
1580 FIELD #2, 3 AS R$,7 AS F$,8 AS Z$,4 AS D$,5 AS L$,5 AS A$,7 AS C$,6 AS G$,6 AS H$,7
AS M$,10 AS Y$,10 AS X$
1590 FOR X=1 TO RN%:GET #2,X
1600 A=VAL(A$):C=VAL(C$):G=VAL(G$):H=VAL(H$):M=VAL(M$)
1610 PRINT:INPUT "Printout?  (Y/N)    ",PR$:PRINT:IF PR$<>"Y" THEN 1620 ELSE 1740
1620 PRINT"Flight description:                      ";F$
1630 PRINT"Starting date:                           ";Z$
1640 PRINT"Distance in nautical miles:              ";D$
1650 PRINT"Fuel consumed:  (Gallons or pounds)      ";L$
1660 PRINT USING "Average fuel cost per unit:              $$##.##";A
1670 PRINT USING "Cost of other consumables:             $$####,.##";C
1680 PRINT USING "Landing fees:                          $$###.##";G
1690 PRINT USING "Cost of tiedown, hangar:               $$###.##";H
1700 PRINT USING "Misc. expenses:                        $$####,.##";M:GOSUB 130
1710 PRINT USING "Total fuel cost:                       $$#####,.##";CVI(Y$)
1720 PRINT USING "Total entire trip:                     $$#####,.##";CVI(X$)
1730 NEXT X:CLOSE #2:GOSUB 130:GOSUB 160:GOSUB 150:GOTO 230
1740 LPRINT"Flight description:                      ";F$
1750 LPRINT"Starting date:                           ";Z$
1760 LPRINT"Distance in nautical miles:              ";D$
1770 LPRINT"Fuel consumed:  (Gallons or pounds)      ";L$
1780 LPRINT USING "Average fuel cost per unit:             $$##.##";A
1790 LPRINT USING "Cost of other consumables:            $$####,.##";C
1800 LPRINT USING "Landing fees:                         $$###.##";G
1810 LPRINT USING "Cost of tiedown, hangar:              $$###.##";H
1820 LPRINT USING "Misc. expenses:                       $$####,.##";M:GOSUB 140
1830 LPRINT USING "Total fuel cost:                      $$#####,.##";CVI(Y$)
1840 LPRINT USING "Total entire trip:                    $$#####,.##";CVI(X$)
1850 NEXT X:CLOSE #2:GOSUB 150:GOTO 230
1860 OPEN "I",#1,"RECORD.NUM":INPUT #1,RN%:CLOSE #1
1870 OPEN "R",#2,"TRIP.DTA",79
1880 FIELD #2, 3 AS R$,7 AS F$,8 AS Z$,4 AS D$,5 AS L$,5 AS A$,7 AS C$,6 AS G$,6 AS H$,7
AS M$,10 AS Y$,10 AS X$
1890 FOR X=1 TO RN%:GET #2,X
1900 INPUT "Printout?  (Y/N)    ",PR$:PRINT:IF PR$<>"Y" THEN 1910 ELSE 1940
1910 PRINT"Flight description:                      ";F$
1920 PRINT"Starting date:                           ";Z$:GOSUB 130
1930 NEXT X:CLOSE#2:GOSUB 160:GOSUB 150:GOTO 230
1940 LPRINT "Flight description:                     ";F$
1950 LPRINT "Starting date:                          ";Z$:GOSUB 140
1960 GOTO 1930
1970 OPEN "I",#3,"MAINT.NUM":INPUT #3,MN%:CLOSE #3
1980 OPEN "R",#4,"MAINT.DTA",44
1990 FIELD #4,3 AS N$,8 AS D$,8 AS F$,8 AS E$,8 AS A$,8 AS O$
2000 FOR X=1 TO MN%:F=VAL(F$):E=VAL(E$):A=VAL(A$):O=VAL(O$)
2010 PRINT:INPUT "Printout?  (Y/N)    ",YN$:PRINT
2020 GET #4,X:IF YN$<>"Y" THEN 2030 ELSE 2160
2030 PRINT"Date:                                    ";D$
2040 PRINT USING "Airframe maintenance expense:          $$#####,.##";F
2050 PRINT USING "Engine maintenance expense:            $$#####,.##";E
2060 PRINT USING "Avionics maintenance expense:          $$#####,.##";A
2070 PRINT USING "Other expense:                         $$#####,.##";O
2080 FF=FF+F:EE=EE+E:AA=AA+A:OO=OO+O:TT=FF+EE+AA+OO:GOSUB 130
2090 PRINT USING "Total Airframe expense to this point:  $$######,.##";FF
2100 PRINT USING "Total Engine expense to this point:    $$######,.##";EE
2110 PRINT USING "Total Avionics expense to this point:  $$######,.##";AA
2120 PRINT USING "Total Other expense to this point:     $$######,.##";OO:GOSUB 130
2130 PRINT USING "Total to this point:                   $$######,.##";TT
2140 GOSUB 130
2150 NEXT X:CLOSE #4:GOSUB 130:GOSUB 160:GOSUB 150:GOTO 230
2160 LPRINT "Date:                                   ";D$
```

continued

```
2170 LPRINT USING "Airframe maintenance expense:          $$#####,.##";F
2180 LPRINT USING "Engine maintenance expense:           $$#####,.##";E
2190 LPRINT USING "Avionics maintenance expense:         $$#####,.##";A
2200 LPRINT USING "Other expense:                        $$#####,.##";O
2210 FF=FF+F:EE=EE+E:AA=AA+A:OO=OO+O:TT=FF+EE+AA+OO:GOSUB 140
2220 LPRINT USING "Total Airframe expense to this point:  $$######,.##";FF
2230 LPRINT USING "Total Engine expense to this point:    $$######,.##";EE
2240 LPRINT USING "Total Avionics expense to this point:  $$######,.##";AA
2250 LPRINT USING "Total other expense to this point:     $$######,.##";OO:GOSUB 140
2260 LPRINT USING "Total to this point:                   $$######,.##";TT
2270 GOSUB 140
2280 GOTO 2150
```

Figure 42.4 The aircraft expense record program.

Lines 100–280 of the program (Figure 42.4) display the title of the program and the main menu, sending the computer to one of three line numbers, depending on the selection made by the user.

Lines 290–340 display the second menu, again sending the computer to one of several line numbers.

Lines 350–370 create a sequential file called RECORD.NUM, used internally by the computer to cause each entry to be numbered consecutively. A random-access file called TRIP.DTA is used to store the trip-related data.

Lines 380–590 ask that the trip-related data be keyed in, computes the totals for each trip, and stores the complete group of data in the random-access file.

Lines 600–730 perform the identical task for the maintenance expense data, creating a sequential file called MAINT.NUM and a random-access file called MAINT.DTA. The sequential file numbers the entries and the random-access file stores the data.

Lines 740–870 display two consecutive menus, sending the computer to line numbers determined by the selections made from those two menus.

Lines 880–1210 are used to display or print the trip data for a flight, based on the starting date of that flight. Lines 1220–1550 perform the identical task for a trip based on a given flight description.

Lines 1560–1850 come into play when the expense data for all flights to date are to be displayed or printed.

Lines 1860–1960 cause just the dates and flight description for all trips to be displayed or printed.

Lines 1970–2280 are used to display or print the previously-entered maintenance data, along with the consecutive subtotals and totals.

CHAPTER FORTY-THREE

Travel Mode Comparison

As long as we own and operate our own aircraft, we'll be tempted to use it for any trip we have to make, giving little thought to the economy of another mode of travel. This program compares the cost and time en route for any trip you may wish to make, using four modes of transportation: your own or your company's airplane, commercial airlines, automobile and bus.

It takes into account such frequently ignored data as the time and cost of travel to and from the airport or bus terminal, the waiting time at the airport, the cost of rental cars at the destination, and so on. When you run the program for more than one of the modes of transportation, it also displays a comparison table, showing the differences in terms of expense and time.

When activated, the program briefly displays a description of its purpose, followed by the main menu:

```
Do you plan to travel by:
---------------------------------------------------------------

  1   Business or private aircraft?
  2   Commercial airlines?
  3   Automobile?
  4   Bus?
---------------------------------------------------------------

  5   Examine comparison data
---------------------------------------------------------------

  6   Exit program
---------------------------------------------------------------

Which?
```

Depending on which of the first four choices is selected, four differing series of prompts will be presented.

For personal or business aircraft:

```
Distance to travel in statute miles?
Average gound speed? (knots)
Fixed costs per hour?                                              $
Cost of fuel per gallon?
Cost of travel to the airport?                                    $
Time of travel to the airport?                                    $
Cost of travel from arrival airport?
Cost of rental car (if any)?                                      $
Food and lodging at destination?                                  $
```

For commercial airlines:

```
Distance to travel in miles?
Scheduled air time (HR.MIN)?
Cost of travel to airport?                                        $
Time of travel to airport?
Waiting time at airport?
Number of paying passengers?
Cost of each ticket, one way?                                     $
Cost of travel from arrival airport?                              $
Time of travel from arrival airport?
Cost of rental car (if any)?                                      $
Food and lodging at destination?                                  $
```

For travel by car:

```
Distance to travel in miles?
Average speed (incl. stops)?
Average cost of gas per gallon?                                   $
Average miles per gallon?
Annual car expenses (excl. gas)?                                  $
Average miles driven annually?
Food and lodging en route?                                        $
Food and lodging at destination?                                  $
```

For travel by bus:

```
Distance to travel in miles?
Scheduled bus travel time (HR.MIN)?
Cost of travel to bus terminal?                                   $
Time of travel to bus terminal?
```

Waiting time at bus terminal?
Number of paying passengers?
Cost of each ticket, one way? $
Food en route? $
Cost of travel from arrival terminal? $
Time of travel from arrival terminal?
Cost of rental car (if any)? $
Food and lodging at destination? $

In each instance, the program uses the input data to compute displayed results:

Grand total (one way): $xxx.xx
Grand total (round trip): $xxx.xx

Final total, entire trip: $xxx.xx

If you want to compare the results for each mode of travel, the program produces this display:

COMPARISON OF COST AND TIME DATA

	Aircraft	Airlines	Automobile	Bus
Trip cost:	$x,xxx.xx	$x,xxx.xx	$x,xxx.xx	$x,xxx.xx
Time 1 way:	xx:xx	xx:xx	xx:xx	xx:xx

```
100 REM TRAVEL MODE COMPARISON
110 REM BASIC-80
120 GOTO 180
130 PRINT STRING$(80,45):RETURN
140 HOME:VTAB(10):RETURN
150 PRINT:INPUT "Press >RETURN< (Q to quit)  ",R$:IF R$="Q" THEN 160 ELSE RETURN
160 GOSUB 140:GOSUB 130:PRINT TAB(38)"End.":GOSUB 130:END
170 PRINT:INPUT "Entries correct?  (Y/N)  ",CORRECT$:RETURN
180 T1$="Approximate time to destination:":T2$=" hours and ":T3$=" minutes."
190 GTO$="Grand total (one way):       $":GTR$="Grand total (round trip)    $"
200 FT$="Final total, entire trip:    $"
210 GOSUB 140:PRINT TAB(22)"TRAVEL MODE COMPARISON PROGRAM"
220 PRINT TAB(22)"*****************************"
230 PRINT:PRINT"This program determines travel costs and times for different modes of"
240 PRINT"transportation: business or private aircraft, airlines, automobile, bus "
250 GOSUB 130:GOSUB 150:GOSUB 140
260 PRINT"Do you plan to travel by:":GOSUB 130
270 PRINT 1,"Business or private aircraft?"
280 PRINT 2,"Commercial airlines?"
290 PRINT 3,"Automobile?"
300 PRINT 4,"Bus?":GOSUB 130
```

continued

```
310 PRINT 5,"Examine comparison data":GOSUB 130
320 PRINT 6,"Exit program":GOSUB 130
330 INPUT "Which? ",WHICH:GOSUB 140
340 ON WHICH GOTO 1150,880,350,590,1400,160
350 INPUT "Distance to travel in miles?          ",MILES
360 INPUT "Average speed (incl. stops)?          ",SPEED
370 INPUT "Average cost of gas per gallon?   $",GAS
380 INPUT "Average miles per gallon?          ",MPG
390 INPUT "Annual car expenses (excl. gas)?  $",YEAR
400 INPUT "Average miles driven annually?       ",ANNUAL:GOSUB 130
410 GOSUB 170:IF CORRECT$<>"N" THEN 420 ELSE GOSUB 140:GOTO 350
420 TIME=MILES/SPEED:TIME1=INT(TIME):TIME2=(TIME-TIME1)*.6*100:TIME2=CINT(TIME2)
430 TIME3=TIME1:TIME4=TIME2
440 PRINT T1$;TIME1;T2$;TIME2;T3$:GOSUB 140
450 GAS=GAS*MILES/MPG
460 PRINT USING "Fuel cost:                           $$#####,.##";GAS
470 YEAR=YEAR/ANNUAL*MILES
480 PRINT USING "Other car costs:                     $$#####,.##";YEAR
490 TOTAL=GAS+YEAR
500 PRINT USING "Total car costs:                     $$#####,.##";TOTAL:GOSUB 130
510 INPUT "Food and lodging en route?               $",FOOD:GOSUB 130
520 TOTAL1=TOTAL+FOOD:TOTAL2=TOTAL1*2
530 PRINT USING "Grand total (one way):               $$#####,.##";TOTAL1
540 PRINT USING "Grand total (round trip):            $$#####,.##";TOTAL2:GOSUB 130
550 INPUT "Food and lodging at destination?         $",MOTEL:GOSUB 130
560 TRIP=TOTAL2+MOTEL:TRIPCAR=TRIP
570 PRINT USING "Final total, entire trip:            $$#####,.##";TRIP
580 GOSUB 130:GOSUB 150:GOSUB 140:GOTO 260
590 INPUT "Distance to travel in miles?             ",MILES
600 INPUT "Scheduled bus travel time? (hr.min)      ",TIMEX
610 INPUT "Cost of travel to bus terminal?        $",TAXI
620 INPUT "Time of travel to bus terminal? (hr.min) ",GROUND
630 INPUT "Waiting time at bus terminal? (hr.min)   ",AIRPT
640 INPUT "Number of paying passengers?             ",PERSONS
650 INPUT "Cost of each ticket, one way?          $",TICKET
660 INPUT "Food en route?                         $",FOOD
670 INPUT "Cost of travel from arrival terminal?  $",TAXI1
680 INPUT "Time of travel from arrival terminal? (hr.min) ",GROUNDX
690 INPUT "Cost of rental car (if any)?           $",CAR:GOSUB 130
700 GOSUB 170:IF CORRECT$<>"N" THEN 710 ELSE GOSUB 140:GOTO 590
710 TIMEX1=INT(TIMEX):TIMEX2=(TIMEX-TIMEX1)/.6:TIMEX2=CINT(TIMEX2)
720 TIMEX=TIMEX1+TIMEX2
730 GROUND1=INT(GROUND):GROUND2=(GROUND-GROUND1)/.6
740 GROUND=GROUND1+GROUND2:GROUNDX1=INT(GROUNDX):GROUNDX2=(GROUNDX-GROUNDX1)/.6
750 GROUNDX=GROUNDX1+GROUNDX2:AIRPT1=INT(AIRPT)
760 AIRPT2=(AIRPT-AIRPT1)/.6:AIRPT=AIRPT1+AIRPT2
770 TIME=TIMEX+GROUND+GROUNDX+AIRPT
780 TIME1=INT(TIME):TIME2=(TIME-TIME1)*.6*100:TIME2=CINT(TIME2)
790 TIME9=TIME1:TIME10=TIME2
800 PRINT T1$;TIME1;T2$;TIME2;T3$:GOSUB 140
810 COST=TAXI+(PERSONS*TICKET)+TAXI1+FOOD:TCOST=COST*2+CAR:TOTAL5=TCOST
820 PRINT USING "Grand total (one way):               $$#####,.##";COST
830 PRINT USING "Grand total (round trip):            $$#####,.##";TCOST:GOSUB 130
840 GOSUB 130:INPUT "Food and lodging at destination?         $",MOTEL:GOSUB 130
850 TRIPBUS=TCOST+MOTEL
860 PRINT USING "Final total, entire trip:            $$#####,.##";TRIPBUS
870 GOSUB 130:GOSUB 150:GOSUB 140:GOTO 260
880 INPUT "Distance to travel in miles?             ",MILES
890 INPUT "Scheduled air time? (hr.min)             ",TIMEX
900 INPUT "Cost of travel to airport?             $",TAXI
910 INPUT "Time of travel to airport? (hr.min)      ",GROUND
920 INPUT "Waiting time at airport?   (hr.min)      ",AIRPT
930 INPUT "Number of paying passengers?             ",PERSONS
940 INPUT "Cost of each ticket, one way?          $",TICKET
```

```
950 INPUT "Cost of travel from arrival airport?          $",TAXI1
960 INPUT "Time of travel from arrival airport? (hr.min) ",GROUNDX
970 INPUT "Cost of rental car (if any)?                   $",CAR:GOSUB 130
980 GOSUB 170:IF CORRECT$<>"N" THEN 990 ELSE GOSUB 140:GOTO 880
990 TIMEX1=INT(TIMEX):TIMEX2=(TIMEX-TIMEX1)/.6:TIMEX=TIMEX1+TIMEX2
1000 GROUND1=INT(GROUND):GROUND2=(GROUND-GROUND1)/.6
1010 GROUND=GROUND1+GROUND2:GROUNDX1=INT(GROUNDX):GROUNDX2=(GROUNDX-GROUNDX1)/.6
1020 GROUNDX=GROUNDX1+GROUNDX2:AIRPT1=INT(AIRPT)
1030 AIRPT2=(AIRPT-AIRPT1)/.6:AIRPT=AIRPT1+AIRPT2
1040 TIME=TIMEX+GROUND+GROUNDX+AIRPT
1050 TIME1=INT(TIME):TIME2=(TIME-TIME1)*.6*100:TIME2=CINT(TIME2)
1060 TIME7=TIME1:TIME8=TIME2
1070 PRINT T1$;TIME1;T2$;TIME2;T3$:GOSUB 140
1080 COST=TAXI+(PERSONS*TICKET)+TAXI1:TCOST=COST*2+CAR:TOTAL5=TCOST
1090 PRINT USING "Grand total (one way):          $$#####,.##";COST
1100 PRINT USING "Grand total (round trip):       $$#####,.##";TCOST:GOSUB 130
1110 GOSUB 130:INPUT "Food and lodging at destination?      $",MOTEL:GOSUB 130
1120 TRIPAL=TCOST+MOTEL
1130 PRINT USING "Final total, entire trip:       $$#####,.##";TRIPAL
1140 GOSUB 130:GOSUB 150:GOSUB 140:GOTO 260
1150 INPUT "Distance to travel in statute miles?        ",MILES
1160 INPUT "Average ground speed? (knots)               ",SPEED
1170 INPUT "Fixed costs per hour?                      $",COST
1180 INPUT "Cost of fuel per gallon?                   $",FUEL
1190 INPUT "Fuel flow per hour? (gph)                   ",FLOW
1200 INPUT "Cost of travel to the airport?             $",TAXI
1210 INPUT "Time of travel to the airport? (hr.min)     ",GROUND
1220 INPUT "Cost of travel from arrival airport?       $",TAXI1
1230 INPUT "Time of travel from arrival airport? (hr.min) ",GROUNDX
1240 INPUT "Cost of rental car (if any)?               $",CAR:GOSUB 130
1250 GOSUB 170:IF CORRECT$<>"N" THEN 1260 ELSE GOSUB 140:GOTO 1150
1260 SPEED=SPEED/1.15089
1270 TIME=MILES/SPEED:COST=(COST*TIME)+(FUEL*FLOW*TIME)+TAXI+TAXI1
1280 GROUNDA=INT(GROUND):GROUNDB=(GROUND-GROUNDA)/.6:GROUND=GROUNDA+GROUNDB
1290 GROUNDC=INT(GROUNDX):GROUNDD=(GROUNDX-GROUNDC)/.6:GROUNDX=GROUNDC+GROUNDD
1300 TIME=TIME+GROUND+GROUNDX:TCOST=COST*2+CAR
1310 TIME1=INT(TIME):TIME2=(TIME-TIME1)*.6*100:TIME2=CINT(TIME2)
1320 TIME5=TIME1:TIME6=TIME2
1330 PRINT T1$;TIME1;T2$;TIME2;T3$:GOSUB 140
1340 PRINT USING "Grand total (one way):          $$#####,.##";COST
1350 PRINT USING "Grand total (round trip):       $$#####,.##";TCOST:GOSUB 130
1360 GOSUB 130:INPUT "Food and lodging at destination?      $",MOTEL:GOSUB 130
1370 TRIPAIR=TCOST+MOTEL
1380 PRINT USING "Final total, entire trip:       $$#####,.##";TRIPAIR
1390 GOSUB 130:GOSUB 150:GOSUB 140:GOTO 260
1400 PRINT TAB(20)"COMPARISON OF COST AND TIME DATA"
1410 PRINT TAB(20)"==================================":PRINT
1420 PRINT TAB(20)"Aircraft";TAB(35)"Airlines";TAB(50)"Automobile";TAB(65)"Bus"
1430 PRINT TAB(20)STRING$(13,45);TAB(35)STRING$(13,45);TAB(50)STRING$(13,45);
1440 PRINT TAB(65)STRING$(13,45):PRINT
1450 PRINT"Trip cost:";
1460 PRINT TAB(20) USING "$$#####,.##";TRIPAIR;
1470 PRINT TAB(35) USING "$$#####,.##";TRIPAL;
1480 PRINT TAB(50) USING "$$#####,.##";TRIPCAR;
1490 PRINT TAB(65) USING "$$#####,.##";TRIPBUS
1500 PRINT:PRINT"Time 1 way:";
1510 PRINT TAB(24)TIME5;":";TIME6;
1520 PRINT TAB(39)TIME7;":";TIME8;
1530 PRINT TAB(53)TIME3;":";TIME4;
1540 PRINT TAB(68)TIME9;":";TIME10:PRINT:GOSUB 130:GOSUB 150:GOSUB 140:GOTO 260
```

Figure 43.1 The travel mode comparison program.

Lines 100–340 of the program (Figure 43.1) display the title and purpose of the program, followed by the menu, sending the computer to one of several line numbers. Lines 350–580 are used for travel by automobile. Lines 590–870 are used for travel by bus. Lines 880–1140 deal with travel by commercial airlines, and lines 1150–1390 summarize travel by business or personal aircraft. Finally, lines 1400–1540 display the comparison table.

CHAPTER FORTY-FOUR

Converting Local Time to Worldwide Time Zones

If you're a corporate pilot flying for an international company, or if you do a lot of traveling to other countries, or, for that matter, if you do a lot of international business on the telephone, it would be handy to know the time of day in Peking, Tokyo, New Delhi, or any other part of the world.

This program performs that task quickly and easily for most major countries and U.S. cities. It starts by asking that you reply to a number of questions:

```
This program converts local time to time anywhere in the world
-------------------------------------------------------------------------
Do you want to convert from local
-------------------------------------------------------------------------
    1   Standard Time?
    2   Daylight Time?
-------------------------------------------------------------------------
Which?


Do you want to convert from:

    1   Eastern Time
    2   Central Time
    3   Mountain Time
    4   Pacific Time
    5   Yukon Time
    6   Alaska/Hawaii Time
    7   Bering Time
-------------------------------------------------------------------------
Which?
```

Do you want to convert to:

```
1  a U.S. City?
2  a foreign country?
```

Which?

It then displays a list of U.S. cities or a list of foreign countries, asking that you make the appropriate selection by number:

U.S. cities (alphabetically):

```
 1  Albuquerque
 2  Anchorage
 3  Atlanta
 4  Boston
 5  Chicago
 6  Cincinnati
 7  Cleveland
 8  Dallas
 9  Denver
10  El Paso
11  Honolulu
12  Houston
13  Los Angeles
14  Miami
15  New Orleans
16  New York
17  Philadelphia
18  Phoenix
19  Portland
20  San Francisco
21  Seattle
22  Washington
```

Which?

or

Countries (alphabetically):

```
 1 Argentina
 2 Brazil
 3 China
 4 England
 5 Egypt
 6 France
 7 Germany
 8 Greece
 9 India
10 Ireland
11 Israel
12 Italy
13 Japan
14 Kenya
15 Philippines
16 Portugal
17 Russia (Moscow area)
18 South Africa
19 Spain
20 Turkey
```

Which?

Once the country or U.S. city has been selected, the program asks for hours and minutes entered separately:

```
Local time? (hour only, no minutes)
Local time? (minutes)
AM (A) or PM (P)?
```

The program then displays the time in the selected country or city and Greenwich Mean Time.

```
100 REM CONVERTING LOCAL TIME TO TIME ZONES WORLDWIDE
110 REM BASIC-80
120 GOTO 170
130 PRINT STRING$(80,45):RETURN
140 HOME:VTAB(10):RETURN
150 PRINT:INPUT "Press >RETURN< ",R$:RETURN
160 GOSUB 140:GOSUB 130:PRINT TAB(38)"End.":GOSUB 130:END
170 T$="The time in the city you asked for is ":GOSUB 140:GOSUB 130
180 TT$="The time in the country you asked for is "
```

continued

```
190 GMT$="Greenwich Mean Time is ":ST$=" Standard Time"
200 PRINT TAB(5)"This program converts local time to time anywhere in the world"
210 GOSUB 130:GOSUB 150:GOSUB 140
220 PRINT"Do you want to convert from local":PRINT
230 PRINT 1,"Standard Time?"
240 PRINT 2,"Daylight Time?":GOSUB 130
250 INPUT "Which?  ",WHICH:GOSUB 140
260 PRINT"Do you want to convert from":PRINT
270 PRINT 1,"Eastern Time":PRINT 2,"Central Time":PRINT 3,"Mountain Time"
280 PRINT 4,"Pacific Time":PRINT 5,"Yukon Time":PRINT 6,"Alaska/Hawaii Time"
290 PRINT 7,"Bering Time":GOSUB 130
300 INPUT "Which?  ",TIME:GOSUB 140
310 PRINT"Do you want to convert to":PRINT
320 PRINT 1,"a U.S. city?"
330 PRINT 2,"a foreign country?":GOSUB 130
340 INPUT "Which?  ",CTY:GOSUB 140
350 ON CTY GOTO 360,590
360 PRINT"U.S. cities (alphabetically):":GOSUB 130
370 PRINT 1,"Albuquerque":PRINT 2,"Anchorage":PRINT 3,"Atlanta":PRINT 4,"Boston"
380 PRINT 5,"Chicago":PRINT 6,"Cincinnati":PRINT 7,"Cleveland":PRINT 8,"Dallas"
390 PRINT 9,"Denver":PRINT 10,"El Paso"
400 PRINT 11,"Honululu":PRINT:INPUT "Choose one of these (C) or go on (G)?  ",CG$
410 IF CG$="C" THEN GOSUB 130:GOTO 450 ELSE PRINT:GOTO 420
420 PRINT 12,"Houston":PRINT 13,"Los Angeles":PRINT 14,"Miami":PRINT 15,"New Orleans"
430 PRINT 16,"New York":PRINT 17,"Philadelphia":PRINT 18,"Phoenix":PRINT 19,"Portland"
440 PRINT 20,"San Francisco":PRINT 21,"Seattle":PRINT 22,"Washington":GOSUB 130
450 INPUT "Which?  ",CITY:GOSUB 140
460 GOSUB 680
470 GOTO 480
480 GOSUB 130:IF CTRY=9 THEN 500 ELSE 490
490 PRINT T$;TIME1;":";MINUTE;" ";M$;ST$:GOTO 510
500 PRINT T$;TIME1;":";MIN;" ";M$;ST$
510 GOSUB 130:GOSUB 530
520 PRINT GMT$;TIME2;":";MINUTE;" ";M$:GOSUB 130:GOSUB 150:GOTO 920
530 TIME2=HOUR+(TIME+4):IF TIME2>12 THEN M$="PM" ELSE M$="AM"
540 IF TIME2>24 THEN M$="AM"
550 IF TIME2>24 THEN TIME2=TIME2-24
560 IF TIME2>12 THEN TIME2=TIME2-12
570 IF TIME2<1 THEN TIME2=TIME2+12
580 RETURN
590 PRINT"Countries (alphabetically):":GOSUB 130
600 PRINT 1,"Argentina":PRINT 2,"Brazil":PRINT 3,"China":PRINT 4,"England":PRINT 5,"Egypt"
610 PRINT 6,"France":PRINT 7,"Germany":PRINT 8,"Greece":PRINT 9,"India":PRINT 10,"Ireland"
620 PRINT 11,"Israel":PRINT 12,"Italy":PRINT 13,"Japan":PRINT 14,"Kenya"
630 PRINT 15,"Philippines":PRINT 16,"Portugal":PRINT 17,"Russia (Moscow area)"
640 PRINT 18,"South Africa":PRINT 19,"Spain":PRINT 20,"Turkey":GOSUB 130
650 INPUT "Which?  ",CTRY:GOSUB 140
660 GOSUB 940
670 T$=TT$:GOTO 480
680 TT=TIME+4:IF WHICH=2 THEN TT=TT+1
690 INPUT "Local time? (hour only, no minutes)          ",HOUR
700 INPUT "Local time? (minutes)                        ",MINUTE
710 INPUT "AM (A) or PM (P)?                            ",AM$
720 IF AM$="P" THEN HOUR=HOUR+12
730 TIME2=HOUR+TT
740 IF CITY=3 OR CITY=4 OR CITY=6 OR CITY=7 OR CITY=14 THEN GMT=TT-5
750 IF CITY=16 OR CITY=17 OR CITY=22 THEN GMT=TT-5
760 IF CITY=5 OR CITY=8 OR CITY=12 OR CITY=15 THEN GMT=TT-6
770 IF CITY=1 OR CITY=9 OR CITY=10 OR CITY=18 THEN GMT=TT-7
780 IF CITY=13 OR CITY=19 OR CITY=20 OR CITY=21 THEN GMT=TT-8
790 IF CITY=2 OR CITY=11 THEN GMT=TT-10
800 TIME1=TIME2+GMT-TT:TIME2=TIME1-GMT+TT
810 IF TIME1>12 THEN M$="PM"
820 IF TIME1<12 THEN M$="AM"
```

```
830 IF TIME1=12 THEN 840 ELSE 850
840 IF MINUTE=0 THEN M$="noon" ELSE M$="PM"
850 IF TIME1=24 THEN 860 ELSE 870
860 IF MINUTE=0 THEN M$="midnight" ELSE M$="AM"
870 IF TIME1>24 THEN M$="AM"
880 IF TIME1>24 THEN TIME1=TIME1-24
890 IF TIME1>12 THEN TIME1=TIME1-12
900 IF TIME1<1 THEN TIME1=TIME1+12
910 RETURN
920 GOSUB 140:INPUT "Another time conversion? (Y/N)   ",TC$
930 IF TC$<>"N" THEN GOSUB 140:GOTO 220 ELSE 160
940 TT=TIME+4:IF WHICH=2 THEN TT=TT+1
950 INPUT "Local time? (hour only, no minutes)      ",HOUR
960 INPUT "Local time? (minutes)                    ",MINUTE
970 INPUT "AM (A) or PM (P)?                         ",AM$
980 IF AM$="P" THEN HOUR=HOUR+12
990 TIME2=HOUR+TT
1000 IF CTRY=4 OR CTRY=10 OR CTRY=16 THEN GMT=TT
1010 IF CTRY=6 OR CTRY=7 OR CTRY=12 OR CTRY=19 OR CTRY=8 THEN GMT=TT+1
1020 IF CTRY=20 OR CTRY=11 OR CTRY=18 OR CTRY=5 THEN GMT=TT+2
1030 IF CTRY=14 OR CTRY=17 THEN GMT=TT+3
1040 IF CTRY=1 OR CTRY=2 THEN GMT=TT-3
1050 IF CTRY=9 THEN GMT=TT+5
1060 IF CTRY=9 THEN MIN=MIN+30
1070 IF MIN>59 THEN MIN=MIN-60
1080 IF MIN>59 THEN GMT=GMT+1
1090 IF CTRY=15 THEN GMT=TT+10
1100 IF CTRY=3 THEN GMT=TT+8
1110 IF CTRY=13 THEN GMT=TT+9
1120 GOTO 800
```

Figure 44.1   The world-wide time-zone conversion program.

Lines 100–190 of the program (Figure 44.1) contain a group of subroutines and assign several strings to string variables. Lines 200–350 ask a number of questions about the conversion to be performed. Lines 360–450 contain the list of U.S. cities. Line 460 sends the computer to a subroutine (lines 680–910) where the conversions are computed. Lines 470–580 display the result, and include another subroutine that makes sure the time is displayed in 12-hour rather than 24-hour format. Lines 590–650 contain the list of foreign countries. Line 660 sends to the computer another subroutine (lines 940–1120, followed by lines 800–910) that performs the calculations for the foreign countries.

CHAPTER FORTY - FIVE

Exchange Rates for Foreign Currencies

This program can be helpful in converting amounts of U.S. dollars to any other currency, or vice versa. All you're asked to do is to key in the amount to be converted and the currently applicable rate of exchange.

To start, the program asks:

Menu:

```
1   Convert FROM U.S. dollars TO a foreign currency
2   Convert FROM a foreign currency TO U.S. dollars
```

Which?

followed by:

Is the currency unit to which you want to convert greater or smaller than the U.S. dollar (i.e. the English Pound)? (G/S)

After this you'll be confronted with either of the following, (the amounts and currency names used here are arbitrary):

Amount of U.S dollars to be converted?

Convert $100 to which foreign currency? (MARKS, YEN, LIRA etc.)
At the current rate of exchange, $1.00 = how many MARKS?

$100 convert to 175.25 MARKS

or

Currency to convert to U.S. dollars? (MARKS, YEN, LIRA etc.)

How many MARKS do you want to convert?
At the current rate of exchange, $1.00 = how many MARKS?

---

100 MARKS convert to $57.14

The display differs slightly from the one shown here if one unit of the foreign currency is greater in value than the U.S. dollar, in which case you're asked:

At the current rate of exchange, 1 POUND = how many U.S. dollars?

```
100 REM CURRENCY EXCHANGE RATES
110 REM BASIC-80
120 GOTO 170
130 PRINT STRING$(80,45):RETURN
140 HOME:VTAB(10):RETURN
150 PRINT:INPUT "Press >RETURN<  (Q to quit)  ",R$:IF R$="Q" THEN 160 ELSE RETURN
160 GOSUB 140:GOSUB 130:PRINT TAB(38)"End.":GOSUB 130:END
170 GOSUB 140:PRINT TAB(26)"CURRENCY EXCHANGE RATES":PRINT TAB(26)"************************"
180 PRINT"Menu:":GOSUB 130
190 PRINT 1,"Convert FROM U.S. dollars TO a foreign currency":PRINT
200 PRINT 2,"Convert FROM a foreign currency TO U.S. dollars":GOSUB 130
210 INPUT "Which?   ",WHICH:GOSUB 140
220 ON WHICH GOTO 230,390
230 PRINT"Is the currency unit to which you want to convert greater or smaller":PRINT
240 INPUT "than the U.S. dollar (i.e. the English Pound)? (G/S)     ",GS$
250 GOSUB 140
260 INPUT "Amount of U.S. dollars to be converted?        $",AMT:PRINT
270 PRINT USING "Convert $$######,.##";AMT;
280 INPUT " to which foreign currency?  (MARKS, YEN, LIRA etc.) ",FC$:PRINT
290 IF GS$="S" THEN 300 ELSE 360
300 PRINT"At the current rate of exchange, $1.00 = how many ";FC$;
310 INPUT EXC:SUM=AMT*EXC:GOSUB 130
320 PRINT USING "$$######,.##";AMT;
330 PRINT" convert to ";
340 PRINT USING "######,.##";SUM;
350 PRINT" ";FC$:GOSUB 130:GOSUB 150:GOTO 380
360 PRINT"At the current rate of exchange, one ";FC$;" = how many U.S. dollars";
370 INPUT EXC:SUM=AMT/EXC:GOSUB 130:GOTO 320
380 PRINT:PRINT:INPUT "Another conversion?  (Y/N)   ",YN$:IF YN$<>"Y" THEN 160 ELSE 170
390 PRINT"Is the currency unit from which you want to convert greater or smaller":PRINT
400 INPUT "than the U.S. dollar (i.e. the English Pound)? (G/S)     ",GS$
410 GOSUB 140
420 INPUT "Currency to convert to U.S. dollars? (MARKS, YEN, LIRA etc.) ",FC$:PRINT
430 PRINT"How many ";FC$;" do you want to convert";
440 INPUT AMT:IF GS$="S" THEN PRINT:GOTO 450 ELSE PRINT:GOTO 470
450 PRINT"At the current rate of exchange, $1.00 = how many ";FC$;
460 INPUT EXC:SUM=AMT/EXC:GOSUB 130:GOTO 490
470 PRINT"At the current rate of exchange, one ";FC$;" = how many U.S. dollars";
480 INPUT EXC:SUM=AMT*EXC:GOSUB 130
490 PRINT USING "######,.##";AMT;
500 PRINT" ";FC$;" convert to ";
510 PRINT USING "$$######,.##";SUM:GOTO 380
```

Figure 45.1 The program that converts U.S. dollars to any foreign currency and vice versa, using prevailing rates of exchange.

As written, the program (Figure 45.1) can accept values of up to 999,999.99 dollars or other currencies. If greater amounts are to be converted, the number of # -signs in front of the comma and decimal point (,.) in lines 340 and 490 must be increased to reflect the maximum number of digits to be used to the left of the decimal point.

Lines 100–220 display the title of the program and the main menu. Lines 230–380 are used if the conversion is to be from U.S. dollars to a foreign currency, and lines 390–510 are used if the conversion is to be from a foreign currency to U.S. dollars.

Converting BASIC-80 to Other Dialects of Basic

Most every computer manufacturer, in addition to the usual BASIC commands, adds a number of commands and statements unique to each system. Depending on the system you're using, certain changes in the programs will have to be made so they will run properly:

CINT(x) in *BASIC-80* produces a rounded-off integer. In *Applesoft and other versions* you must use **X = INT(X + .5)** for a rounded-off integer, and **X = INT(X * 100 + .5)/100** to limit the number of decimals to the number of zeros used (two in this example).

FILE CONVENTIONS differ greatly among computers. *For programs that create their own sequential or random-access files*, be sure to check the instruction manual for your computer. The differences are too enormous to be described here.

FOR PAUSE = 1 TO X:NEXT PAUSE, available in *BASIC-80, Applesoft and most other dialects*, creates a pause of X length before program execution is continued. Using **FOR A = 1 TO X:NEXT A** produces the same result.

HOME is a command available in the *Apple version of BASIC-80 and in Applesoft*. It clears the screen and places the cursor on the left top corner of the screen. The TRS-80s and IBM PCs use **CLS** to achieve the same result. The TI-99/4A uses **CALL CLEAR**, which clears the screen but leaves the cursor in the Bottom left corner of the screen. To move it up, use a **FOR...TO...NEXT** loop, with the number after TO representing the number of lines up you want the cursor to move.

IF...THEN...ELSE is a statement in which the ELSE portion is not available in *Applesoft and some other systems*. Instead of using ELSE, simply go to the next line.

INPUT "prompt statement" is used in several versions of BASIC, but in APPLESOFT it must be followed by a **semicolon**, and in TI BASIC it must be followed by a **colon**. Other computers, such as the ATARI and TRS-80 require that the prompt statement be used as a **PRINT** *line preceding the INPUT command*.

LEFT$, MID$, RIGHT$ are statements available in *BASIC-80, Applesoft and some other dialects of BASIC* for the purpose of assigning portions of a string to new string variables. TI BASIC uses **SEG$** instead, and ATARI uses any string variable followed by one or two subscripts in parentheses.

LINE PRINTER COMMANDS cause data to be sent to the line printer. In BASIC-80 the command is **LPRINT**, and the same convention is used by the TRS-80, IBM PC and ATARI. TI BASIC requires **OPEN #1: "RS232"** followed by **PRINT #1: "TEXT"**, where TEXT is whatever you want printed. After printing is finished, a **CLOSE #1** statement is required. *Applesoft* uses **PR# 1** (assuming the printer interface card is in slot #1) to activate the printer, after which all PRINT material will be sent to the line printer. To stop, type **PR#0**.

MULTIPLE STATEMENTS ON ONE LINE are separated by a **single colon** in *Applesoft and most other dialects*, but by **two colons (::)** in TI EXTENDED BASIC. TI BASIC does not accept more than one statement per line.

ON ERROR GOTO must be changed to **ONERR GOTO** in *Applesoft and certain other dialects*.

PRINT USING "prompt statement $$ # # # # #,. # #";X is a statement available in *BASIC-80* but not in *Applesoft and most other versions*. It's purpose is to format the manner in which numberic data are displayed or printed. In the above example, 10000 would be displayed as $10,000.00. It can be replaced by simple PRINT statements, in which case leading and trailing zeros will not be displayed. Or, it can be replaced by a subroutine that converts numeric variables to string variables. In the following example the numeric variable in question is Z:

```
100 Z = INT(Z*100 + .5)/100 : ZZ = INT(Z) : IF ZZ = Z THEN 170
110 IF Z<1 THEN 130
120 GOTO 140
130 Y$ = "0"
140 ZZ = *10 : Z1 = INT(ZZ) : Z2 = ZZ – Z1 : IF Z2 < .1 THEN Z$ = "0"
150 IF Z<1 THEN 190
160 GOTO 180
170 Z$ = ".00"
180 NN$ = STR$(Z) + Z$ : RETURN
190 NN$ = Y$ + STR$(Z) + Z$ : RETURN
```

NN$ is the string variable that will print the number including leading and/or trailing zeros. **LPRINT USING** is the same as above, except that the formatted data are sent to the line printer.

RANDOMIZE and **RND** are the statements that produce a random number. Some BASIC versions require a "seed number" in conjunction with RANDOMIZE, others do not, and still others do not use RANDOMIZE at all. Check your manual.

TAB(x) requires a **semicolon** between it and the next statement in *TI BASIC and some other versions*.

VARIABLE NAMES may be of any length in *BASIC-80, BASICA and TI BASIC*, with 40 or more characters significant. The significant characters are **limited to two** (plus $ in case of string variables) in *Applesoft and many other versions of BASIC*.

VTAB(x) moves the cursor x lines down from the top. It is not available in *many versions of BASIC* and can be replaced with **FOR A=1 TO x:PRINT:NEXT A**, which has the same effect.

When converting a program in this book to another dialect of Basic, and you're not certain which commands and statements might have to be changed, the simplest routine might be to type the program, as listed, into your computer. Then try to run it. The program will run until the computer encounters an unacceptable expression. On most systems this will produce a "Syntax Error On Line xxx" message, telling you where to look for the command or statement that must be changed. Once it has been corrected, repeat the procedure to find the next error, and so on.

If the program contains an ON ERROR GOTO statement, you must temporarily delete it during this process, as *any* error, including syntax errors, will cause the computer to go to the indicated line number instead of alerting you to the statement or command that produces the error message.

Glossary of Aviation Terms and Abbreviations

absolute ceiling   The maximum altitude above sea level to which a particular aircraft can climb and then maintain horizontal flight under standard atmospheric conditions.

accelerate-stop distance   The distance required to accelerate an aircraft from standing start to lift-off speed. Assuming failure of the (critical) engine at the instant that speed is reached, the distance to bring the aircraft to a stop using heavy breaking.

ADF   Automatic direction finder.

ADF approach   A non-precision instrument approach using NDBs for lateral guidance.

aerodynamics   The forces, such as resistance, pressure, velocity and others involved in the movement of air or gasses around a moving body. Conversely, the branch of dynamics and physics dealing with these forces.

agl   Above ground level.

AH   Artificial horizon.

ailerons   The primary control surfaces located at the trailing edges of the outer wing panels which, when deflected up or down, cause the airplane in flight to bank.

air-data instruments	The instruments used by the pilot in the control of the aircraft, such as airspeed indicator, altimeter, turn-and-bank indicator and so on.
airfoil	Any surface designed to create lift, either positive or negative, when moving through the air at a given speed. Primarily wings and control surfaces, though propellers and helicopter blades are also airfoils.
airspace	When used in aviation, the term means the navigable airspace. For all practical purposes, between ground level and 60,000 feet.
airspeed	The speed with which an aircraft is moving with relation to the air around it. It may be expressed as indicated airspeed, calibrated airspeed or true airspeed.
airspeed indicator	A flight instrument with a cockpit read-out which, in terms of knots or mph, shows the difference between pitot pressure and static pressure. The reading obtained from the airspeed indicator is indicated airspeed.
alternator	An electrical device, serving the same purpose as the old-style generator, driven by the engine. It supplies electrical current to the battery and to all on-board electrical equipment except the ignition system.
altimeter	A flight instrument capable of displaying the height above sea level (or any other predetermined level), activated by an aneroid barometer measuring atmospheric pressure at a given altitude.
altimeter setting	The barometric pressure reading in the small window provided for that purpose on the face of the altimeter.
ambiguity meter	The TO/FROM indicator in an OBI.
angle of attack	The angle at which the chord line of the wing or other airfoil meets the relative wind. It determines the amount of lift developed at a given airspeed.

approach	The maneuvers an airplane needs to perform prior to landing.
Approach control	The ATC facility monitoring and directing traffic approaching an airport where such a facility is in operation.
area navigation	Navigating along direct routes rather than airways, using special equipment capable of electronically relocating DME-equipped nav aids, or VLF/Omega or INS long-range navigation equipment.
ARTCC	Air Route Traffic Control Center, usually simply referred to as "center". The ATC facilities handling en route IFR traffic.
artificial horizon	A gyro instrument showing the altitude of the aircraft with reference to pitch and roll as compared to the horizon.
ATC	Air Traffic Control.
atmospheric pressure	The weight of the air surrounding the earth in layers of varying characteristics. Standard atmospheric pressure is expressed as 29.92 inches of mercury, or 1613.2 millibars.
avionics	A catch-all phrase for communication, navigation and related instrumentation in an aircraft. A contraction of "aviation electronics."
back-course approach	A non-precision instrument approach in the direction opposite the localizer, using the localizer for lateral guidance.
balanced field length	The distance within which a jet aircraft can accelerate to VI and then stop or accelerate to a safe climb speed, V2, and clear a 35-foot obstacle on one engine.
base leg	A part of the airport traffic pattern. A flight path at right angles to the runway, following the down wind leg and followed by the final approach.
C.	Celsius or centigrade.

calibrated airspeed	Indicated airspeed corrected for instrument and installation errors.
CAS	Calibrated airspeed.
carburetor heat	A heating unit located near the carburetor throat and controlled by a plunger in the cockpit. It is used to melt or prevent carburetor ice.
carburetor ice	Ice forming in the carburetor throat due to excessive amounts of moisture in the air.
CDI	Course deviation indicator.
com	Communication (radios).
compass, gyro	A compass system driven by a gyroscope, not reacting to the magnetic field of the earth.
compass, magnetic	A compass which, during straight and level flight, automatically aligns itself with magnetic north. It is unreliable during turns, climbs and descents.
compass rose	The compass card showing 360 degrees.
controllable-pitch propeller	See constant-speed propeller.
controlled airport	Airports at which a control tower is in operation.
controlled airspace	Those portions of the airspace in which aircraft must operate under ATC control.
constant-speed propeller	A controllable-pitch propeller that maintains a constant rpm by automatically changing the blade angle in relation to engine power output.
course	The direction of flight of an aircraft across the ground. Also see Heading.
course deviation indicator	The needle, bar or other indicator in an OBI that displays the position of an aircraft relative to a radial or bearing from or to a VOR.
CRS	Course.

dead reckoning	A method of navigation. The course and time of an aircraft between two given points is estimated by taking course, speed and wind components into consideration, calculated with a wind triangle. The phrase "dead reckoning" is a bastardization of the term "deduced reckoning."
decision height	The lowest altitude to which an aircraft may descend during a precision instrument approach without visual contact with the ground.
ded reckoning	Dead reckoning.
density altitude	Pressure altitude corrected for prevailing temperature conditions.
Departure Control	The ATC facility controlling departures at airports at which such service is available.
detonation	The burning of the fuel-air mixture by explosion rather than steady burning. It results in rapidly rising cylinder-head temperatures and can result in catastrophic engine damage.
dew point	The temperature to which air must cool in order for condensation to take place without change in pressure or vapor content.
DG	Directional gyro (compass).
DH	Decision height.
directional gyro	A gyroscopic flight instrument which, when set to conform with the magnetic compass, will continue to indicate the aircraft heading for some time, regardless of turns or pitch changes. It tends to develop heading errors and must be adjusted intermittantly.
distance measuring equipment	A combination of airborne and ground equipment that gives a reading of the distance from the airplane to a ground station by measuring the time lapse of a signal generated in the airplane and reflected by the ground station. Most systems also display ground speed and time to station, using the rapidity of change in the distance data.

DME	Distance measuring equipment.
downwind	Away from the direction from which the wind is blowing.
downwind leg	The flight path parallel to the runway in the direction opposite landing. It is part of the standard airport traffic pattern.
drag	The force created by friction of the air on subjects in motion. It must be overcome by thrust to achieve flight parallel to the relative wind. There are two types of drag, "induced" drag and "parasite" drag. Induced drag is created through the process of creating lift. Parasite drag is all drag from surfaces that do not contribute to lift. It increases with an increase in airspeed.
EGT	Exhaust gas temperature (gauge).
elevator	The primary control surface, attached to the horizontal stabilizer, that can be deflected up or down to control the pitch of the aircraft. It is, in fact, primarily a speed control, not an altitude control.
encoding altimeter	An instrument that senses the current altimeter reading and transmits this information to the transponder which, in turn, automatically transmits it to ATC.
engine analyzer	An exhaust gas temperature gauge with probes and read-outs for each cylinder of the engine.
E6b	A circular slide rule-type computer used to compute a variety of flight-related mathematical problems.
exhaust gas temperature	The temperature of the gases escaping through the exhaust manifold. The temperature of these gases varys with the fuel-air mixture being used, and it can be measured and displayed in the cockpit through the use of an exhaust gas temperature gauge or an engine analyzer.
F.	Fahrenheit.

FAA	Federal Aviation Administration.
fanjet	An aircraft driven by a turbine engine, combining the effects of direct jet thrust with the thrust produced by an engine driven shrouded fan.
FBO	Fixed base operator.
final approach	The final portion of an airport traffic pattern (or the final leg of a straight-in approach), during which the aircraft is aligned with the runway center line.
fixed base operator	Service operations located on an airport.
fixed-pitch propeller	A propeller, the blade angle of which cannot be changed or adjusted.
FL	Flight level. FL180 stands for 18,000 feet.
flaps	Auxiliary control surfaces, usually located at the trailing edges of the inner wing panels between the fuselage and the ailerons. Flaps can be extended and/or turned down to increase the wing camber and/or surface, creating additional lift or drag.
flare	A smooth leveling of the aircraft during which the nose is raised at the end of the glide and just prior to touch-down.
flight path angle	The angle between a horizontal line and the flight path of an aircraft during climb or descent, measured in degrees.
flight plan	A flight profile filed with the FAA prior to flight. VFR flight plans are voluntary. IFR flight plans are mandatory.
flight service station	A facility, usually located on an airport, where weather information is available and pilots may file flight plans.
FPA	Flight path angle
fpm	Feet per minute in terms of climb or descent.
FSS	Flight service station.

ft	Feet.
GCA	Ground controlled approach. A non-precision approach during with the ground-based controller uses radar to provide lateral guidance for the pilot.
generator	A device, identical in construction to an electric motor, that, generates electrical current and continuously charges the battery when driven by an engine.
glide slope	A ground-based nav aid that provides the pilot with a cockpit display in the appropriately equipped OBI, showing the aircraft's position relative to the prescribed glide path.
gph	Gallons per hour (fuel flow).
gradient	As used on instrument approach charts, the altitude change in feet over a distance of one nautical mile.
Greenwich Mean Time	The standard time at Greenwich, England, used in aviation worldwide to avoid confusion caused by different time zones. Also referred to as "Zulu time."
Ground Control	An ATC service at controlled airports, responsible for the safe and efficient movement of aircraft and airport vehicles on the ground.
ground effect	A certain amount of additional lift which takes effect when the aircraft is close to the ground. It is the result of air compression between the wings and the ground. Low-wing aircraft are more susceptible to the effect than high-wing aircraft.
ground speed	The speed with which the aircraft moves across the ground. Not to be confused with airspeed.
GS	Ground speed or glide slope.
HDG	Heading.
heading	The direction in which the aircraft flies through the air, not with reference to the ground. In other words, the direction in which the nose of the aircraft is pointing. Not to be confused with course.

horizontal stabilizer	The fixed horizontal section of the empennage to which the elevator is attached.
IAS	Indicated airspeed.
IFR	Instrument flight rules.
ILS	Instrument landing system.
ILS approach	A precision instrument approach using the localizer for lateral, and the glide slope for vertical, guidance.
IM	Inner marker.
inches of mercury	A unit of measure of atmospheric pressure, indicating the height in inches to which a column of mercury will rise in a glass tube in response to the weight of the atmosphere exerting pressure on a bowl of mercury at the base of that tube.
indicated airspeed	The airspeed displayed by the airspeed indicator. It is nearly always less than true airspeed, but usually not much different from calibrated airspeed.
induced drag	See drag.
in hg	Inches of mercury.
INS	Inertial navigation system. A long-range navigation system operating independent of any ground-based navigation aids, at sea level.
intersection take-off	A take-off using less than the full length of the runway.
ISA	ICAO Standard Atmosphere. Standard atmospheric conditions, such as 15 degrees C. or 59 degrees F. at sea level.
isogonic lines	The lines on aeronautical charts indicating magnetic variations.
kHz	Kilohertz or kilocycles.
knots	Nautical miles per hour.
kts	Knots.

laminar flow   Air flowing smoothly over and adhering to the surface of an airfoil.

lift   The generally upward force created by the difference of pressure between the upper and lower surfaces of an airfoil in motion. In level flight it is balanced by the force of gravity.

LOC   Localizer.

localizer   A portion of an instrument landing system that provides horizontal guidance to the pilot.

localizer approach A non-precision instrument approach using the localizer without glide slope.

magnetic course   The course of an aircraft referenced to the magnetic north.

magnetic heading   The heading of an aircraft referenced to the magnetic north.

magnetic north   The region, some distance from the geographic north pole, where the earth's magnetic lines concentrate.

magnetic variation The angle between magnetic and true norths. It differs at various points on the earth and is shown as isogonic lines on aviation charts.

magneto   A self-contained generator that supplies electrical current to the spark plugs in the ignition system.

manifold   An arrangement of tubing with one orifice on one end and several on the other.

manifold pressure   The pressure of the fuel-air mixture in the intake manifold.

MAP   Missed approach point.

MDA   Minimum descent altitude.

MEA   Minimum en route altitude.

MHz   MegaHertz or megacycles.

mHz   MiniHertz or minicycles.

minimum descent altitude	The altitude to which any aircraft may descend without visual contact with the ground during a non-precision approach.
minimum en route altitude	The minimum safe altitude during instrument flight. It varys and is dependent on the ability to receive signals from the appropriate ground-based navigation aids.
minimum obstruction clearance altitude	The minimum safe altitude during instrument flight based on ground elevation and obstructions with no regard to the nav aid reception distances.
minimum reception altitude	The minimum altitude at which the aircraft can receive one of the two nav aids, representing either end of one leg of an airway.
missed approach point	The point during an instrument approach at which a missed approach must be initiated if there is no visual contact with the ground.
mixture	The mixture of fuel and air necessary for combustion in reciprocating engines.
MLS	Microwave landing system.
msl	Height in feet above mean sea level.
MM	Middle marker.
MOA	Military operations area.
MOCA	Minimum obstruction clearance altitude.
monocoque	Type of all-metal aircraft construction in which the fuselage shell carries most of the structural loads.
MP	Manifold pressure.
mph	Statute miles per hour.
MRA	Minimum reception altitude.
mushing	Flying along in a nose high attitude at low speeds.
NASA	National Aviation and Space Administration.
nav	Navigation (radio).

NDB	Non-directional beacon.
needle and ball	An instrument that shows the degree of bank and displays whether the aircraft is in a skid or slip. An older version of the modern turn-and-bank indicator.
nm	Nautical mile(s).
non-precision approach	An instrument approach using nav aids or radar for lateral but not for vertical guidance.
normally aspirated	A reciprocating engine that is not turbocharged or mechanically supercharged.
OAT	Outside air temperature.
OBI	Omni-bearing indicator.
OBS	Omni bearing selector.
OM	Outer marker.
omni-bearing indicator	The cockpit display that shows the horizontal relationship of the aircraft with reference to a selected VOR radial or bearing. If equipped with a glide-slope needle it also provides vertical guidance during a precision approach.
omni-bearing selector	The knob on the OBI used to select a given VOR.
PAR	Precision approach radar (approach). A ground-controlled approach in which the controller uses radar that displays the position of the aircraft horzontally and vertically.
parasite drag	See drag.
pattern	Airport landing pattern; the downwind leg, base leg and final approach.
peak	Expression used to denote the highest possible exhaust gas temperature.

phonetic alphabet

The official alphabet used in aviation radio communication:

A	Alpha	**M**	Mike	**Y**	Yankee
B	Bravo	**N**	November	**Z**	Zulu
C	Charlie	**O**	Oscar	**0**	Zero
D	Delta	**P**	Papa	**1**	Wun
E	Echo	**Q**	Quebec	**2**	Too
F	Foxtrot	**R**	Romeo	**3**	Tree
G	Gulf	**S**	Sierra	**4**	Fow-er
H	Hotel	**T**	Tango	**5**	Fife
I	India	**U**	Uniform	**6**	Six
J	Juliett	**V**	Victor	**7**	Sev-en
K	Kilo	**W**	Whiskey	**8**	Ait
L	Lima	**X**	X-ray	**9**	Nin-er

pitch

The attitude of the aircraft with reference to the horizontal axis at right angles to the fuselage. In other words, nose down or nose up.

pilotage

Navigation by reference to visible landmarks. Used usually with sectional charts on which all meaningful landmarks are shown.

pitot-static system

A device that compares pitot pressure with the static or atmospheric pressure and presents the result in the cockpit by means of an airspeed indicator, altimeter and vertical-speed indicator.

pph

Pounds per hour (fuel flow).

precession

The tendency of a directional gyro to become gradually unreliable due to friction.

precision approach

An instrument approach using ground-based nav aids or radar for lateral as well as vertical guidance.

preignition

The burning of the fuel-air mixture in the combustion chamber before the spark plugs have had an opportunity to fire. It usually follows excessive overheating of the engine and can result in serious engine damage.

propeller	A device consisting of two or more airfoil-shaped blades, designed to convert the turning force of the engine into thrust.
propjet	A jet aircraft in which turbine engines drive propellers.
psi	Pressure in terms of pounds per square inch.
relative wind	The movement of air relative to the movement of an airfoil. It is parallel to and in the opposite direction of the flight path of an airplane.
RNAV	Area navigation.
rpm	Revolutions per minute.
rudder	The primary control surface attached to the vertical stabilizer, deflection of which causes the tail of the aircraft to swing left or right. It controls yaw.
runup	A pre-take-off check of the performance of the engine and, in aircraft equipped with constant-speed propellers, the operation of the propeller.
runway visual range	The horizontal visibility at a given runway measured in feet.
RVR	Runway visual range.
SAR	Surveillance approach radar (approach). See GCA.
scan	The consecutive order in which the pilot checks the instrument readings in the cockpit.
sectional chart	An aeronautical chart of a section of the U.S. at a scale of 500,000 to 1, or approximately seven nm to the inch.
service ceiling	The highest altitude at which a given aircraft can continue to climb at 100 fpm.
skid	Lateral movement of an airplane toward the outside of a turn, caused by incorrect rudder use.
slip	The tendency of the aircraft to lose altitude by slipping toward the center of a turn as a result of incorrect rudder use.

sm	Statute mile(s).
special VFR	Conditions below normal VFR minimums under which pilots at controlled airports may take off or land by following simple instructions from the tower. Special VFR conditions are one-mile visibility with no ceiling minimums.
spin	A maneuver in which the airplane, after stalling, descends nearly vertically, nose low, with the tail revolving around the vertical axis of the descent.
spoilers	Flat vertical surfaces that can be raised out of the upper or lower surface of the wing at the pilot's discretion. They spoil the airflow and cause the wing to lose lift. Spoilers may be constructed to be deployed simultaneously in both wings, or individually in one or the other wing. In the first instance they act as speed brakes. In the latter they take the place of ailerons.
stabilator	A primary control surface where the entire horizontal tail surface is movable.
stall	The inability of the aircraft to continue flight due to an excessive angle of attack. It will either drop its nose and thus reduce the angle of attack and regain flying speed, or, if forced to retain the excessive angle of attack, it may fall into a spin.
stall speed	The speed at a given angle of attack at which airflow separation begins and the stall occurs. Aircraft can stall at virtually any speed if an acceptable angle of attack is exceeded.
stall-spin	The combination of stall and spin, a major cause of fatal accidents.
stall warning	A device, usually involving a buzzer, a light or both, that indicates the aircraft is about to stall.
static vent	A hole, usually located in the side of the fuselage, that provides air at atmospheric pressure to operate the pitot-static system.

stick	Control wheel or yoke.
S/VFR	Special VFR.
TAS	True airspeed.
taxi	Moving an aircraft on the ground under its own power.
TBO	Time between major overhauls, expressed in hours of operation.
TCA	Terminal control area. Areas around major air terminals where all aircraft must operate under ATC control.
thrust	The forward force, pulling or pushing, exerted by the engine-driven propeller or, in the case of sailplanes, by gravity. It opposes and must overcome drag.
torque	The normal tendency of an aircraft to rotate to the left in reaction to the right hand rotation of the propeller and the fact that the action of the propeller forces air against the left side of the vertical stabilizer with greater force than against the right side. It alters with changes in power.
tower	Control tower at controlled airports.
transceiver	An instrument capable of receiving as well as transmitting voice or other signals.
transponder	An airborne radar beacon transceiver that automatically transmits responses to interrogation by ground-based transmitters.
trim tab	A small airfoil attached to control surfaces that can be adjusted to cause changes in the position of the control surface under varying flight conditions, reducing pilot work load.
true airspeed	The actual speed at which the aircraft is moving through the air. It is calibrated airspeed adjusted for the prevailing air density and altitude.

true course	Course referenced to true north.
true heading	Heading referenced to true north.
T-tail	The tail section of an airplane where the horizontal stabilizer or stabilator is attached to the top of the vertical stabilizer.
turbo-charging	A process in which a turbine, driven by the exhaust gases, compresses the air and increases the amount of fuel-air mixture available to the engine. Used primarily to increase the high-altitude performance of piston-engine aircraft.
turbojet	A jet aircraft driven by turbines without the use of propellers or fans.
turboprop	See propjet.
turn-and-bank indicator	See needle and ball.
uncontrolled airport	An airport without a control tower, or where the control tower is not operating.
unicom	Aeronautical advisory station for communication with aircraft (also the associated frequencies). Usually manned by FBOs, unicoms are not authorized to give clearances except when relaying word from ATC, in which case transmissions must be preceded by the words "ATC clears. . ."
unusual attitude	Any attitude of an aircraft in terms of pitch, roll or both, that is beyond the normal operating attitude.
venturi	A tube that is narrower in the center than at either end. When air is forced through a venturi it results in suction that can be used to drive gyro instruments in aircraft not equipped with vacuum pumps.
vertical speed indicator	An instrument, part of the pitot-static system, that indicates the rate of climb or descent in terms of fpm.
vertical stabilizer	A fixed vertical airfoil on the empennage to which the rudder is attached.

VFR	Visual flight rules.
VHF	Very high frequency: electromagnetic frequencies between 30 and 300 MHz.
Victor Airways	Designated airways, located between VORs, VOR/DMEs or VORTACs.
VLF	Very low frequency; below 30 kHz.
VLF/Omega	A long-range navigation system using the world-wide network of VLF and Omega stations for lateral guidance.
VOR	Very high frequency omni-directional radio range; a ground-based VHF navigation aid.
VOR approach	A non-precision instrument approach using VORs for lateral guidance.
VOR/DME	A VOR with DME capability.
VORTAC	Another type of VOR with DME capability.
VSI	Vertical speed indicator.
WAC	World Aeronautical Chart; scale 1,000,000 to 1.
wake turbulence	Turbulence created by movement of an aircraft through the air. Primarily the trailing wingtip vortices that develop in the wake of heavy aircraft. Resembling a pair of counter-rotation horizontal tornadoes, they are at their worst behind slow-moving, heavy jet aircraft just after take-off and immediately prior to landing.
waypoint	A navigational fix used in area navigation. It is created by electronically relocating a VOR/DME or VORTAC from its actual position to a phantom position desired by the pilot.
wind shear	An abrupt change in wind direction or velocity.
wing leveler	A simple autopilot without directional control capability.

yaw	The movement of an aircraft to either side, turning around its vertical axis without banking.
yoke	Control wheel, stick.
Zulu time	Greenwich Mean Time.

Glossary of Computer Terms and Abbreviations

accumulator	Another word for memory or register, derived from the fact that subtotals are accumulated and retained in memory.
address	The phrase or line number used to recall previously stored data from the computer memory.
address register	The register in the CPU that contains the address of the program in use.
ADP	Automatic data processing; the activity of the computer.
Algol	A high-level computer language best suited to process mathematical problems. The acronym: ALGOrithmic Language.
alphanumeric	A combination of alphabetic and numeric data.
analog	One of two methods of solving problems electronically. In the analog method electrical signals are of variable voltages, and the slightest change may cause a significant difference in output. The word "analog" is derived from *analogy*. The alternate method of electronic problem solving is referred to as *digital*.
analog computer	A computer that uses current fluctuations. Once predominent, analog computers have become rare since the invention of the microprocessor.

ANSI	American National Standards Institute, an organization that studies computer languages and codes in an effort to bring about standardization.
AOS	Arithmetic operating system: the conventional method of solving mathematical problems. See also RPN.
APL	A high-level programming language. The acronym stands for A Programming Language.
applications program	A program designed to solve a specific type of problem. All the programs in this book are applications programs. Contrast to systems programs, such as CP/M, MS-DOS etc.
array	A group of variables stored under a single variable name with numeric subscripts.
ASCII	American Standard Code for Information Interchange. It is the code that represents the assembly language and machine code substitutes for upper- and lowercase letters, numbers, punctuation marks, symbols and so on. It is the standard used by most microcomputers.
assembler	A program that translates assembly language into machine code. Used in conjunction with compilers.
assignment statement	A statement that assigns a value, numeric or string, to a variable name.
asterisk (*)	The multiplication sign used in computer programs.
BASIC	Beginners' All-purpose Symbolic Instruction Code: the most popular high-level computer language. There are many different dialects of Basic, such as BASIC-80. The programs in Part Three in this book were written in BASIC-80.
baud	The speed with which information and data are transmitted. A baud rate of 300 refers to a transmission speed of 300 bits per second.

binary	Computers can use only two digits, 0 and 1, representing the state of the electrical circuits (0 = off, 1 = on). All data entered into the computer must be translated into series of zeros and ones. For example, the binary equivalent of the decimal numbers from zero–16 are:

0 00000	**5** 00101	**9** 01001	**13** 01101
1 00001	**6** 00110	**10** 01010	**14** 01110
2 00010	**7** 00111	**11** 01011	**15** 01111
3 00011	**8** 01000	**12** 01100	**16** 10000
4 00100			

bit	The smallest possible unit of information. One bit tells the difference between zero and one.
block diagram	A graphic representation of a program, also referred to as "flowchart."
boot	Stands for "bootstraps," derived from pulling yourself up by your own bootstrap. A short program that prepares the computer to accept data from disk or to write data to disk. To "cold boot" a disk means turning the computer on after inserting the disk in the drive. To "warm boot" means to insert the disk with the computer turned on and then using certain keystroke commands to cause the disk to boot.
branch	A detour in a program, activated in Basic and certain other languages by the statement GOTO and line number or other address.
buffer	A temporary storage area for data, limited by the number of K bytes in the computer's RAM or external buffer.
byte	Any unit of information consisting of eight bits, such as any single letter, symbol, etc.
C	A programming language, distinguished by the speed with which it solves problems.
calculator mode	Refers to using the computer to solve problems instantly, rather than through the use of a program.

cathode-ray tube (CRT)	The video display of the monitor.
character	Any letter, digit, symbol, etc.
chip	An integrated circuit, or the tiny piece of silicon that makes up part of an integrated circuit.
Cobol	A high-level computer language particularly suited to solving business-related problems, such as handling random-access files. The acronym stands for COmmon Business Oriented Language.
column	Each character space on a program or text line.
command	Any phrase in a programming language that causes the computer to perform a given function.
compiler	A program that translates the data, information, statements and commands entered in one of the programming languages into assembly language.
computer error	There is no such thing. It usually refers to human error, though mechanical or electrical malfunctions are, of course, possible.
computer language	The languages used in program writing designed to be compiled or interpreted to make them understandable by the computer.
conditional statement	A statement that tells the computer to perform a given function only if a certain condition exists, such as, in Basic: IF X = 0 THEN GOTO...
constant	A numeric or alphanumeric expression that remains unchanged throughout the program. The opposite of a variable.
cps	Characters per second; the speed with which line printers print.
CPU	Central Processing Unit, the heart of the computer.
CRT	Cathode ray tube.
cursor	The usually flickering mark that appears on the video display to indicate where the next keyed-in character will appear.

data	Any type of information.
DATA	A statement in Basic that permits the entry of any number of alphanumeric data in programs. They can then be accessed through the use of the READ statement.
database program	A program designed to accumulate data of one kind or another, and from which these stored data may be retrieved in random fashion.
data field	An area, usually of predetermined size, into which data can be entered in a random-access file program.
data file	A file created automatically by a program for the purpose of data storage. Data files may be sequential files or random-access files.
data link	A means of electronically transferring data via telephone lines, radio waves, microwaves, coaxial cables, fiber optics, or laser beams.
data processing	The manipulation of data by the computer.
debugging	Editing or correcting a program.
decimal	The fractions represented by the digits to the right of the decimal point.
decimal system	The conventional arithmetic system, using numbers from zero to nine.
default	A condition preprogrammed by the manufacturer of a piece of hardware or software. In most instances the default conditions can be changed by the user.
dialect	A version of a programming language, appropriate for use with a specific computer make or model.
digital	Using numbers rather than analog-type representations in the manipulation and display of data.
digital computer	Computers that operate by counting the ON/OFF sequences in electrical currents. All microcomputers are digital computers.

disk	A medium for data storage. Disks may be hard disks or so-called floppy disks. The latter are available in three sizes: eight inch, 5 1/4-inch and 3.5 inch. They may be single-sided or double-sided, and single-density or double-density.
diskette	Usually refers to a 5 1/4-inch disk.
disk operating system	The system that controls the operation of the disk drives, such as CP/M, DOS 3.3, MS-DOS, etc.
DOS	Disk operating system (pronounced doss).
duplex	A communication system that permits simultaneous transmission and reception, such as an ordinary telephone.
edit mode	A means of editing program lines without retyping the entire line.
EDP	Electronic data processing.
error message	Messages displayed by the computer when a program or syntax error has been detected.
executive program	A program designed to handle the management of the computer system.
expression	Usually refers to an arithmetic expression, such as $2 \times (8 - 3)$.
file	The storage area in memory or on disk where programs, data, text or other information is retained. Files must be given unique file names to permit retrieval of the stored material.
fixed-length record file	Files in which each record consists of a fixed number of bytes.
floppy disk	See disk.
flowchart	See block program.
formatting	To prepare a disk for use. All disks must be formatted in a manner acceptable to the disk operating system with which they are to be used.

Fortran	A high-level programming language best suited to algebraic problems, allowing exponentiation of up to three subscripts. The acronym stands for FORmula TRANslation.
garbage	Useless or incorrect information.
GIGO	Stands for Garbage In, Garbage Out. Incorrect input will result in faulty output.
global search	Looking for a given piece of information anywhere within a program or file.
GOSUB	A Basic command that sends the computer to a subroutine. Must be used in conjunction with RETURN.
GOTO	A Basic command that sends the computer to a branch.
hard copy	Print-outs produced by the line printer.
hardware	All actual equipment, including peripherals, constituting the computer system.
hexadecimal	An arithmetic system using 16 characters:

0 0	**5** 5	**9** 9	**13** D
1 1	**6** 6	**10** A	**14** E
2 2	**7** 7	**11** B	**15** F
3 3	**8** 8	**12** C	**16** 10

high-level language	All programming languages other than assembly language and machine code. Also referred to as procedure-oriented languages (POLS). Languages more comprehensible to the human operators than the alphanumerical representations used by low-level languages.
integer	Any number stripped of its decimals. The integer of 1234.5678 is 1234.
interactive	A program that includes questions and prompts that must be acted upon by the user.

interface	Any device that connects two pieces of hardware.
interpreter	A machine code program that understands, interprets, and executes programs written in a high-level language, usually Basic. Programs written in an interpreted language run slower than those that are compiled.
I/O	Input/Output.
K	Short for kilo. In the area of computers, one K stands for 1024 bytes.
kilobaud	1000 baud.
line printer	Dot-matrix or daisy-wheel printers operated by the computer.
local search	Searching for a piece of information within a limited portion of the program.
loop	A means of repeating a number of program steps a specified number of times. In Basic: FOR X = 1 TO 10: PRINT X:NEXT X, which causes the numbers from 1 to 10 to be printed on successive lines.
M	Mega, or 1,000,000. One megabyte equals one million bytes.
machine language	The only language the computer understands, consisting entirely of zeros and ones. A low-level language.
memory	All those cubbyholes in which the computer stores data, information, text, etc. Computers are equipped with two types of memory, read-only memory (ROM) and random-access memory (RAM). The former is preprogrammed by the manufacturer and cannot be accessed or changed by the user. The latter is the one into which the user enters data for subsequent retrieval.
microcomputer	A personal computer. The expression refers to the fact that the computer uses microprocessor technology, and has nothing to do with size.

microprocessor	The heart and guts of the computer, a sophisticated version of an integrated circuit that is capable of performing a wide variety of computer functions.
microsecond	One millionth of a second.
millisecond	One thousandth of a second.
modem	A piece of hardware needed for computers to communicate with one another over the telephone. Modem stands for Modulate/Demodulate. In conjunction with certain software programs, they modulate computer data into sounds, similar to those generated by push-button telephones, demodulating them on the other end into the computer commands.
nonvolatile memory	A memory capable of retaining stored data even when the computer (or calculator) is shut off.
octal	An arithmetic system using only eight digits:

0	00	**5**	05	**9**	11	**13**	15
1	01	**6**	06	**10**	12	**14**	16
2	02	**7**	07	**11**	13	**15**	17
3	03	**8**	10	**12**	14	**16**	20
4	04						

output	Opposite of input.
overflow	Results when an arithmetic expression results in numbers consisting of more digits than the computer can handle.
page	A screen of information.
Pascal	A structured high-level programming language, named after the French scientist Blaise Pascal.
peripheral	Any hardware device connected to the computer, such as printers, disk drives, etc.
POL	Procedure-oriented language.
procedure	A subroutine or integrated program segment.

procedure-oriented language	See high-level language.
program	A sequence of instructions designed to perform a series of predetermined functions.
RAM	Random-access memory.
random-access file	A file that permits the retrieval of data in any order, eliminating the need to search through the entire file; and one that permits repeated data entry during subsequent runs of the program.
random-access memory	See memory.
READ	See DATA.
read-only memory	See memory.
record	A group of data fields in a data file.
register	A specific location within RAM.
REM	A Basic statement denoting remark lines ignored by the computer.
reserved word	A word that is part of the vocabulary of a programming language and may not be used as a variable.
RETURN	See GOSUB.
reverse Polish notation	Named after the Polish mathematician Jan Lukasiewicz, it is an arithmetic system that does not support equal signs or parentheses. Many of the better pocket calculators use this system.
round-off error	The cumulative error resulting when fractions are rounded-off in mathematical expressions.
RPN	Reverse Polish Notation.

screen-oriented program	A program that displays all instructions and prompts, obviating the need for printed documentation.
sequential file	A file in which data are stored, and from which data are retrieved, in sequential order.
simplex	A communication transmission system that permits either transmission or reception, but not both at the same time.
software	Computer programs.
string	Any sequence of alphanumeric characters. Digits may be included but cannot be used in mathematical calculations.
subroutine	A separate segment of a program that can be accessed repeatedly during a program run.
syntax	The rules that govern the manner in which the computer accepts data in a given programming language.
system	The computer and all peripherals.
terminal	The peripheral to which the computer outputs data, usually the monitor or printer.
toggle key	A key or key combination that reverses a given condition each time it is pressed.
variable	The alphanumeric term to which numeric or string data may be assigned: AA = 10 or AA\$ = "string".
volatile memory	A memory that loses all stored data when the electrical current is shut off. Most microcomputers have volatile memories.
window	A portion of the display that is reserved for input data when other portions are occupied by fixed data, help menus or the like.
word	The number of bits the computer can deal with at one time, usually either eight or 16.

Size of the Do-It-Yourself Programs

File Name	Program Identification	Bytes	Lines
AV.1	Flight data (12 programs)	7,908	170
AV.2	Weight and balance	8,053	156
AV.3	Take-off distance data	4,323	92
AV.4	Range and endurance	4,238	104
AV.5	Altitude selection	9,698	215
AV.6	Flight route file program	6,208	135
AV.7	Great Circle navigation	4,088	86
AV.8	Direct route by one VOR	3,993	81
AV.9	Density altitude	1,521	37
AV.10	Zulu time conversion	1,696	51
AV.11	Aircraft ownership justification	8,916	194
AV.12	Aircraft expense records	12,564	219
AV.13	Travel mode comparison	8,319	145
AV.14	Time zones wold-wide	4,581	103
AV.15	Currency conversion	2,077	42

Software
Manufacturers

Manufacturer	Program	Price
Briley Software P.O. Box 2913 Livermore, CA 94550-0291	RNAV3 NAVIGATOR	$ 29.95
Compuflight Operations Service, Inc. 48 Harbor Park Drive Port Washington, NY 11050		
John T. Dow 6560 Rosemoor Street Pittsburgh, PA 15217	DOW-4 GAZELLE on disk:	$30.00 $33.00
Flight Opps 3857 Birch Street #237 Newport Beach, CA 92660	FLIGHT UTILITIES FLIGHT PLANNER WEATHER STATION ENGINE TREND MONITORING	$95.00 $750.00 $750.00 $250.00
FlightWare Av. Software 16799 Schoenborn Street Sepulveda, CA 91343	NAVLOG	$124.95

G-WHIZ	G-WHIZ FLIGHT PLANNING	$100.00

Michael Falter Int'l.
35 Hamilton Terrace
London NW8 9RG England

Jerry Kennedy Pro Computer Center Route 8 Sweets Corner Plaza Carbondale, IL 62901	FLIGHT PLANNER	$59.95
Mind Systems Corp. P.O. Box 506 Northampton, MA 01061	AIRSIM-3 SPITFIRE SIMULATOR	$44.95 $40.00
PHH Aviation Systems 1726 Cole Blvd. #200 Golden, CO 80401	AvPLANNER AvNAV AvMGR DISPATCH II	$750.00 $95.00 $995.00 to $6,000.00
Ranchele Micro 1010 Shipman Lane McLean, VA 22101	RANCHELE MICRO FLIGHT PLAN	$119.95
Space-Time Assoc. 20-39 Country Club Dr. Rochester, NH 03102	AIR NAV WORKSHOP	$40.00
SubLogic Corp. P.O. Box 4019 Champaign, IL 61820	FLIGHT SIMULATOR II	$49.95

CHAPTER

Index

Accelerate-stop distance, 155–159
Aircraft expense data, 199–214
Airport traffic area, 93–94
Air taxi program, 140
Altitude, cruising, 160, 168
Ambiguity meter, 45
Angle of attack, 81–84
Aviation department software, 134–139,
  144, 145

Balanced field length, 155–159
BASIC-80, 146–148

Center of gravity, 149–154
Champaign VOR, 42
Clearance delivery, 54
Corporate aviation programs, 131–139, 144,
  145
Cruising altitude, 160–168
Currency exchange rates, 226–228

Density altitude, 155–159, 193–195
Direct route, 189–192

Empire State Building, 18
Endurance, 174–177
Expense data, aircraft, 199–214

Fixed Base Operator program, 140
Flight planning programs, 102–109, 114,
  115, 122–133

Flight reference card, 14–15
Flight route data, 178–183

Great circle distances, 184–188
Greenwich Mean Time, 196–198
Ground run, 155–159

Instrument approach, 61
Instrument panel, 3–6
Instrument training programs, 116, 121, 141
Intermediate longitudes, 184–188

LaGuardia Airport, 16, 47
Landing pattern, 59
Landing weight, 149–154
Lindbergh International Airport, 77
Logan International Airport, 47–57
Los Angeles International Airport, 75
Low-altitude en route charts, 28, 37–41

Manhattan, flight over, 16–20
MBASIC, 146–148
Meigs Field, 11–12
Moment arm, 149–154

Obstacle clearance distance, 155–159

Preflight planning programs, 102–109, 114,
  115, 122–133

Ramp weight, 149–154
Range, 174–177

Sectional charts, 28–36

Takeoff weight, 149–154
Terminal control area (TCA), 16
Time zones, 221–225
Travel times, cost, 215–220

Van Nuys Airport, 73, 78–80
Views from the aircraft, 6–9

Willard Field, 42
World Trade Center, 18

Zero-fuel weight, 149–154